MINERAL HANDBOOK
Rapid-Easy Mineral-Rock Determination

Written for Anyone Interested
in Minerals and Rocks

Color Illustrated

By
Paul Dean Proctor
P. Robert Peterson
Uwe Kackstaetter

Macmillan Publishing Company
New York

Collier Macmillan Canada
Toronto

Maxwell Macmillan International
New York Oxford Singapore Sydney

Macmillan Publishing Company
866 Third Avenue, New York, New York 10022

Collier Macmillan Canada, Inc.
1200 Eglinton Avenue, F.
Suite 200
Don Mills, Ontario M3C 3N1

ISBN 0-02-396845-1

Printing: 12345678   Year: 123456789

PREFACE

Prepared for the novice, rockhound, and with sufficient details included to be of interest and use to the field geologist, this handbook should be of great value to anyone interested in minerals and rocks. The emphasis is on mineral identifications through use of easily recognized mineral physical properties and simple chemical tests. Numerous identification tables are provided, such as specific gravity, streak, fusibility, fluorescence, general solubility in waters and acids, and microchemical tests. The section entitled "Tables for the Determination of Non-metallic and Metallic Minerals" includes five identifiers and should be effective in almost 90 percent of the cases if the instructions are carefully followed. Each of these various aids should greatly assist the newly interested to well-trained person to the specific identification of over 300 minerals.

We describe most of the physical and chemical properties of each mineral in the descriptive section. Also included in this section is a color photographic section of each of the described minerals. While an alphabetical listing of the minerals would first appear most easily used, we avoided the easy path and stayed with the more generally used format of chemical classification of the minerals by elements, oxides, carbonates, and other chemical relationships. This has the advantage of keeping related minerals together. It will make it much easier to remember the similar properties of many mineral groups.

To further assist you, some basic descriptions of the underlying principles of crystal chemistry are included. While this is not essential in the identification of a specific mineral, it does give better background information to help you understand why minerals assume certain forms and shapes and retain specific physical properties. If the section appears a bit sophisticated, skip it. You can readily use the other information in your identifications without it. Each of these sections and other descriptive sections are included in the handbook in order to open doors for you to new views of the marvelous world of mineralogy. If your interest is aroused in these new panoramas, reading of suggested references will also expand and enlarge your understanding and appreciation of this new world of knowledge.

Because minerals in aggregate make up all the rocks of the earth, and they are the source of minerals and vice versa, it is appropriate that a brief description of rock genesis, or development, and simple rock classifications of the three major classes of rocks, i.e., igneous, sedimentary and metamorphic, be included. For that most fortunate person who may be privileged to find and identify an extraterrestrial rock material, a brief section is devoted to meteorites and the controversial tektites. Good color photographs illustrate the various rock groups, together with specific descriptions of them.

We end this preface with special thanks to the many who assisted in the preparation of the handbook. Dale Claflin, expert draftsman, and numerous students in the ore minerals laboratory at Brigham Young University successfully used the charts and tables and added many new minerals to their memories. Linda and Brent Davis' expertise in word processing was a major impetus in completing the book. David Lewis and Jana Walker contributed greatly to the mineral and rock color photography. We are deeply indebted for the generosity and help of the Department of Geology at Brigham Young University and for access to the mineral and rock collections. Without such assistance the book could not have been completed. Professor Richard R. Kennedy critically read the manuscript, for which we are most grateful.

To you, the reader, we wish you well and a bon voyage into the kingdoms of minerals and rocks.

TABLE OF CONTENTS

LIST OF FIGURES AND TABLES

FIGURES

TABLES

INTRODUCTION

History has been made by minerals and rocks, the subject of this handbook. From the beginning of man's historical record to now, minerals and metals have been an attraction for him. Abraham of ancient times dealt in gold and silver. Indeed, after discussion , he purchased the burial plot for his wife, Sarah, with 400 shekels (about 200 ounces) of silver. Precious stones and gold were important to King Solomon not only for his own aggrandizement, but more for the great temple at Jerusalem. Gold from his mines was a mainstay in paying for his armies, building projects and the great temple. Searches through the years failed to identify the source of his enormous wealth. Recently an Iranian entrepreneur suggested Africa as the gold source, but a former well-known geologist of the U.S. Geological Survey thinks otherwise. From work in northwest Saudi Arabia, he concludes that the enormous reserves of gold entrapped in the minerals of ancient mine workings and unmined rock are strong indications that this area is the likely source of King Solomon's gold of old.

The localization of certain large metalliferous mining districts within the boundaries of nations is an obvious benefit to that nation so endowed. The United States, for example, led the industrial world for many years because of its abundant and easily accessible mineral resources. With the depletion of many of these deposits such as the hematitic iron ore, the Soviet Union is stepping into the breach and assuming a similar strong leadership role based on its mineral resources. Brazil and Australia now vie for their place in the sun because of an abundant endowment of strategic mineral resources. Japan, a leading industrial nation, gained its position from importation from the Pacific land belt of essential raw materials by inexpensive water routes for the iron, manganese, coal, copper, petroleum, and other mineral sources necessary to its industrial base. This goal of leadership was sought by conquest less than a half century ago, and lost. It has since been won through trade and entrepreneurship in mineral leases, options, purchases, and trades throughout the world.

Minerals and their accumulation into workable economic deposits called ore deposits have affected the course of human history many times. Yet, the aggregate land area of the world involved in these deposits is but a very small fraction of one percent. Mineral deposits, unlike agricultural products, are non-replenishable. A second crop cannot be expected. Rarely an exception occurs as in the case of recovery of salts in evaporating ponds, and their replenishment by the addition of saline waters. For metallic minerals of strategic importance to the nation, the law of no second harvest applies.

The United States is endowed with numerous good grade industrial mineral deposits made up of the compounds of iron, tungsten, molybdenum, vanadium, copper, aluminum, lead and zinc and others. The industrial revolution of our nation coincided with the development of these deposits. The great iron ore region of the Lake Superior district, together with water routes via the Great Lakes, and sources of coal in nearby Illinois, Indiana, Iowa and Pennsylvania, plus an abundance of good metallurgical limestone and fluorspar deposits in the same region were basic to the establishment of the world's greatest steel complex for many years. Only recently, with the depletion of high-grade iron ore reserves, and the discovery and development of other major iron ore districts in the world, has the strength of our steel production begun to decline.

Near Bauxite, Arkansas high-grade deposits of bauxite sustained a major aluminum industry in the U.S. for many years prior to World War II. The deposits of this region were depleted and the United States became dependent on foreign sources with resultant deficits added to our imbalance in trade dollars.

The United States has been much blessed with abundant coal resources. Likewise copper and molybdenum are plentiful even though we do import copper because of lower world prices compared to its own relatively high cost of production. It is dependent on foreign suppliers for manganese with sources in South Africa, Brazil, and Russia. Chrome is derived mainly from South Africa and Turkey, tin from Malaysia and Bolivia, mercury from Spain, nickel from Canada and New Caledonia, and platinum from South Africa. A large new mine in Montana may soon supply part of the U.S. needs of this latter strategic mineral.

As in the present, so in the past, the great silver-lead deposits of Laurium, Greece supplied the monies to fund the navies of the Greeks who, in turn, defeated the Persians in 480 B.C. Jason and his argonauts set sail for the Black Sea and the northern coasts of Turkey to find the Golden Fleece. These were skins of fleece placed in ancient sluice boxes which entrapped the fine gold from the gravels. Dried in the sun, the gold was recovered from the Golden Fleece of ancient times.

Phoenicians of early times sailed to the southwest coast of present England over almost uncharted seas to obtain the tin from Cornwall. In turn, this precious metal was transported back to the eastern Mediterranean and sold at high profits to the people there. Trojan and Grecian soldiers wore armor of bronze derived from the copper and tin of far-away Britain.

From nations to individuals, search and discovery are two of the great joys of life. In the physical world in which we live are many such opportunities for enjoyment. Each day our lives are touched directly by something from the mineral kingdom of the earth. The carbon in the pencil with which we write, the cement of the sidewalk on which we walk, the asphalt of the roads on which we ride, the metal of the cars we drive, and even the glasses we use to improve our sight, all have their common origin in the earth. The list is almost endless.

Have you ever considered what is a rock? Or what is the most abundant rock in the crust of the earth? Or knowing that rocks are made of minerals, what are these substances? Is one more abundant than all the others? Did you know that the all important life-sustaining soils of the earth are made of a mixture of minerals, or that the mountains and their bold outcrops of rock are but a mixture of billions of small mineral substances? Have you considered that some of these were derived from earlier minerals, that others formed from a molten mass called a magma? Consider for a moment minerals and their physical and chemical composition. Minerals are made of atoms of different kinds. The minerals have a generally defined chemical composition, and in almost all cases, a very orderly atomic geometry and symmetry following the laws of nature.

Not many persons can name, much less identify, the eight minerals which constitute 98 percent or more of the earth's crust. And of these few minerals how many of these persons can describe the chemical composition, or the atomic arrangement of these important building blocks of nature?

This handbook will teach you how to recognize and understand the makeup of minerals, not only of the important eight, but many, many more. With thought, persistence and some effort on your part you will search and discover a new kingdom of the earth. Its doors will be opened and you will gaze into one of order and beauty and be able to explore it to whatever extent you desire. We will give you the guidelines and pathways to follow, but only you can tread these paths and enjoy the marvelous facets of the mineral and rock kingdom.

The mineral feldspar, one of the most abundant minerals of the earth's crust, will become a familiar object in this kingdom. You will discover quartz and its many forms and colors, mica with its shining luster, perfect reflecting cleavage planes, and its rare elastic characteristics. The other five minerals will become familiar entities as you proceed along the paths of the mineral kingdom. Like good friends who gather together, you will recognize many mineral friends and their associates. Imagine the joy and satisfaction of being at home anywhere in the world, a true citizen of the world, for the minerals and rocks you will learn to know have no political boundaries. Granite is granite, quartz is quartz, mica always mica, whether you be in Australia, Antarctica, Asia, Africa, or the United States. All such minerals can and will become well known companions wherever you may be in the world.

This handbook, carefully studied, and with the correct use of its numerous and helpful charts and tables, will lead most persons with an interest in minerals to the identification of at least three hundred different mineral species with very little difficulty. The book is arranged to permit ready recognition of minerals in the field or in the laboratory. The student or hobbyist will discover that the charts and the brief mineralogical descriptions clearly summarize the major physical and some definitive chemical characteristics of the minerals. The individual mineral descriptions and accompanying color photographs give details of the physical and chemical properties and the common appearance of the mineral in nature. Through review of the introductory material and the use of the charts, tables and individual mineral descriptions and illustrations, the identification of the minerals will be enjoyable and much simplified. The time requirements will decrease for any specific mineral determination, and the repeated use of the identifiers and mineral descriptions will finally result in memory recognition of the minerals on sight.

Also included in the book is a Systematic Microchemical Analysis procedure of Short and Chace (1940,1951). This useful system has been much neglected by the determinative mineralogist or mineral hobbyist. With the investment of a modest sum in a suggested microchemical kit either by direct purchase or by personally assembling one of your own making, you will have a very powerful tool for the identification of many of the described minerals. The technique is also useful in the determination of upward of 30 specific elements found in the minerals as well. A description and plan for the kit is provided in that section of the book dealing with chemical tests.

This book is arranged to give you experience in the laboratory and field identification methods for minerals and rocks. By applying the scientific method of investigation, you will soon be able to recognize the physical properties of the minerals and make correct identifications.

The unique and easily used charts along with the mineral descriptions will prove most useful in your search. In addition simple chemical confirmatory test are suggested to confirm your findings and make your discoveries a pleasurable experience.

For those whose interest is expanded, many sophisticated techniques are presently available beyond the simple physical and chemical tests we suggest. Among the more refined methods are X-ray Diffraction (XRD) and X-ray Fluorescence (XRF) which are used to determine internal atomic structure and elemental composition. Scanning electron microcopy is an effective method of determining elemental composition and crystalline structure. The Reflecting Microscope (RM) and the Polarizing Microscope (PM) of the mineralogist are very useful instruments in determining the optical and reflective properties of minerals. For determination of elemental composition the emission spectrograph, atomic absorption, and newly developed plasma (ICP), analytical techniques are now available.

In developing this book every attempt has been made to keep it simple and concise in the identification of the more than 300 minerals described.

We begin with a description of the six crystal systems within which all minerals fall. This is followed by concise definition and description of the major physical properties of minerals. These are the simple concepts of MOHS hardness, color, luster, streak, cleavage tenacity, and specific gravity, as well as others.

The identification charts comprise one of the most important sections of the book. Each chart is based on five very easily recognized physical properties of minerals, i.e., color, then cleavage or lack of it, hardness, specific gravity, and luster. If these five simple identifiers can be recognized for a single unknown mineral, the identification of more than 300 minerals described is a distinct possibility. For those relatively rare minerals which have five similar physical properties, it will be necessary to refer to the mineral description section and possibly use some chemical tests in order to make the specific determination.

If the observed physical properties are inadequate for a diagnostic determination, then additional procedures will be needed. Reference to the simple chemical tests provided with the individual mineral descriptions will normally permit specific identifications to be made. Because of the large number of minerals in nature, most of which are very rare, there will be cases in which the minerals are not included in the charts or descriptions. In such cases sophisticated laboratory procedures may be necessary, or the services of a professional mineralogist may be required. This course of action is recommended if the mineral may have economic significance. Please note that the majority of the charts have adequate space to add other minerals as the need arises. Descriptions for such minerals can be found in the more advanced text books of mineralogy.

The color photographs in the descriptive section of the book reflect the "garden variety" rather than the classic "museum variety" of minerals. We felt this approach would be the most useful in assisting you in the identification of minerals in the field.

MINERAL CRYSTAL SYSTEMS

Before describing crystal systems, let us consider the crystal habit of minerals. Only a very few minerals develop into well-formed crystals which come near to resembling the idealized crystal or crystals of the crystallographer. However, many minerals do occur in forms or have a habit of crystallization quite unique to that mineral. Some of the characteristic forms of crystals are sketched in figure 1. The scale of the crystals is usually small, and is measured in millimeters or fractions of an inch; rarely large scale forms may appear as in veins and other openings in rocks and in pegmatites, where conditions for growth are favorable. Mica, for example, has a typical "micaceous" or "foliated" appearance at the small scale of mica grains in an igneous or metamorphic rock, yet masses a foot or more in diameter occur in pegmatites – an igneous rock characterized by large grain size. Other specific minerals almost always crystallize in a radiating form such as pectolite (white cover photograph) and wavellite, while others form fibrous, fiber-like aggregates such as chrysotile (asbestos) and satinspar (gypsum). The typical habit of the mineral or mineral aggregate is included in the mineral description section of the handbook. These habits or forms are an expression of the crystal system of the mineral and are very useful aids in identifying the mineral substance.

Many minerals occur in nature with definite and characteristic external geometric forms. These we call crystals. The outward shapes are an expression of an orderly internal atomic arrangement of the elements which compose the crystals. The possible number of crystal patterns is limited by geometric consideration. As a result, all crystals can be classified within one of six possible crystal systems.

Mineral crystals may occur in single forms, or combinations of more than one form within its specific crystal system. In some cases twin crystals may develop of the same mineral and these are intergrown according to definite laws of crystallography (figure 10).

 Micaceous or foliated

 Botryoidal, grape-like cluster

 Fine granular, earthy, aggregate of grains

 Reniform, somewhat rounded or kidney-shaped

 Scaly, made up of scales

 Prismatic, or elongate crystals with parallel faces

Tubular, flat, platelike

 Acicular, needle-like

 Columnar, columnar groups

 Radiate, radiating from common point

 Bladed, crystals flat like a knife blade

Figure 1   Some common crystal forms and crystal aggregates of minerals.

The Crystal Systems

The six crystal systems are isometric, hexagonal, tetragonal, orthorhombic, monoclinic, and triclinic. The system names are indicative of the characteristics of imaginary crystal axes which lie within the crystals. We measure three axes for descriptive purposes of man made objects such as length, width, and depth. In crystal nomenclature three imaginary crystal axes are also established.

In the **isometric system** three axes are at right angles to each other and are of equal length (iso=equal, metron=measure). These are used by crystallographers to number and name crystal faces in accordance with the intersection which a face or faces make with one or more of the axes of the crystal. A cubic face, for example, cuts one axis at right angles and parallels the other two, and is numbered 100 (1=cuts; 0=parallel). The last of the three digits refers to the vertical imaginary axis. Hence, a 001 face would parallel the two horizontal axes and cut the vertical one. A 111 octahedral face would have eight possible repeats and cut the edge of intersecting cubic faces equally and axes at the same distance from their common point in space. Some crystal faces may not visibly cut an axis. The trapezohedron of figure 2 cuts the horizontal axis if extended right, but it is twice as far away as for regular intersecting axes and crystal faces. The farthest intersection is labelled 1 (1=cut), the other number reflects the fraction of the distance needed to make a regular cut (2=cut at 1/2), and is numbered 211. Proportional distances 3, 4, or 5 times greater have equivalent fractios to reflect the distances. The well-known mineral galena commonly develops cubic, octahedral, and even dodecahedral (12 faces) in its crystallization habit (figure 2).

The isometric crystal system has many forms and shapes and high order symmetry. It must always have, however, three equal length imaginary axes at right angles to each other. Some crystal forms of the isometric system are shown in figure 2. Three imaginary crystal axes are sketched in all but one of the crystals. Visualize where they would be in this rhombododecohedron model.

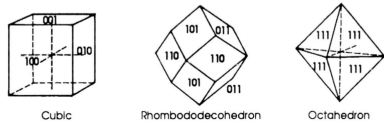

| Cubic | Rhombododecohedron | Octahedron |

Some typical isometric minerals: diamond, gold, platinum, copper, argentite, chromite, magnetite, fluorite, sphalerite galena, halite, tetrahedrite, and others.

| Trapezohedron | Pyritohedron | Tetrahedron |

Some other isometric minerals: garnet, spinel, cuprite, pyrite, bornite, cobaltite, uraninite and others with extended crystal face intersections.

Figure 2. Examples of the isometric crystal system. Mineral forms and common mineral axes in the system shown by dashed lines.

The **Tetragonal System.** This crystal system (Tetra=four, gonon=angle) has three imaginary crystal axes at right angles to each other. The two horizontal axes are of equal length and the vertical axis is longer or shorter than the other two. Zircon, for example, crystallizes in the tetragonal system. A common crystal form is a 4-sided pyramid which may have vertical sides or prismatic faces, and an upside down pyramid (below). Figure 3 shows three crystal forms in the tetragonal system and gives a short list of some common tetragonal minerals.

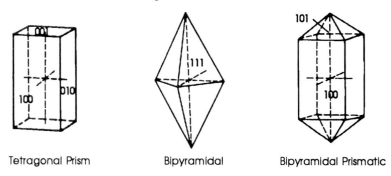

Tetragonal Prism   Bipyramidal   Bipyramidal Prismatic

Common tetragonal minerals include: cassiterite, chalcopyrite, rutile, scheelite, thorite, idocrase, wulfenite, zircon and others.

Figure 3.   Some crystal forms in the tetragonal crystal system.

The **Hexagonal System** (hexa=six, gonon=angle) has four crystal axes. Three of the axes are equal in length and in a horizontal plane. These make angles of 60 degrees with each other. The fourth axis is of different length and at right angles to the others, and in the vertical plane of the crystal. The common mineral quartz shows the distinctive characteristics of the hexagonal system (figure 4).

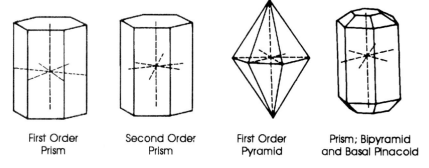

First Order   Second Order   First Order   Prism; Bipyramid
Prism   Prism   Pyramid   and Basal Pinacoid

Common hexagonal minerals are: apatite, beryl, molybdenite, niccolite, pyrrhotite, quartz and zincite.

Figure 4.   Hexagonal crystal forms with crystal axes.

Note that hexagonal first order and second order prisms differ. In first order prisms three equal length axes intersect the crystal faces at their edges. In second order

8

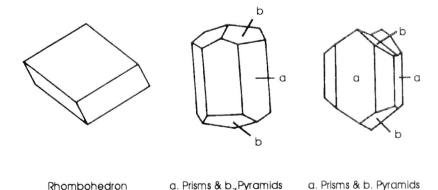

| Rhombohedron | a. Prisms & b. Pyramids | a. Prisms & b. Pyramids |

Some common minerals in the class: chabazite, brucite, beryl, alunite, calcite, dolomite, tourmaline, rhodochrosite, siderite, corundum, hematite, cinnabar and others.

Figure 5. Examples of crystals of the rhombohedral class of the hexagonal system.

The **Orthorhombic System** (orthos=straight) has three imaginary crystal axes at right angles to each other. Each has a different length. The mineral barite crystallizes in the system. Figure 6 shows some typical crystal forms in the system and includes a list of some common minerals in it.

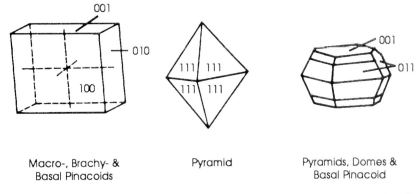

| Macro-, Brachy- & Basal Pinacoids | Pyramid | Pyramids, Domes & Basal Pinacoid |

Common minerals in the system include: aragonite, azurite, barite, brochantite, cerussite, columbite, enargite, staurolite, stibnite, strontianite, marcasite, olivine, topaz and others.

Figure 6. Some crystal forms in the orthorhombic crystal system.

The **Monoclinic System** (mono=single, cline=inclined) has three crystal axes of unequal length, two of which intersect obliquely and at right angles to the third. The gypsum crystal exhibits these characteristics as well as crystal forms of biotite and muscovite micas. Common crystal forms and minerals in the system are shown in figure 7.

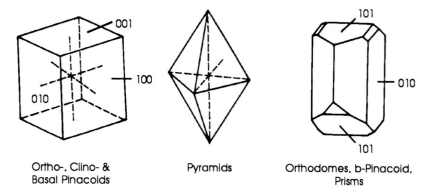

Ortho-, Clino- & Basal Pinacoids	Pyramids	Orthodomes, b-Pinacoid, Prisms

Augite, biotite, crocoite, gypsum, hornblende, malachite, monazite, muscovite, orthoclase, tenorite, titanite, wolframite and numerous others are examples.

Figure7. Common mineral forms in the monoclinic system.

The **Triclinic System** (tri=three, cline=inclined) has three unequal crystal axes. Each has a mutually oblique intersection with the other two. Kyanite, plagioclase, and rhodonite crystallize within this crystal system. Some crystal forms in the triclinic system are shown in figure 8 together with some common minerals.

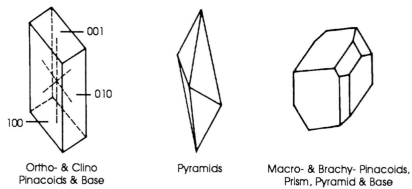

Ortho- & Clino Pinacoids & Base	Pyramids	Macro- & Brachy- Pinacoids, Prism, Pyramid & Base

Plagioclase, kyanite, rhodonite, turquoise, ulexite and axinite are common minerals in system.

Figure 8. Some crystal forms in the triclinic system.

The crystal forms illustrated are only a few of the very large number possible and known in nature. Yet, these can be very useful guides as you study the various natural crystals. The minerals pyrite and garnet and their various crystal forms illustrate some of the numerous possibilities. Pyrite may develop cubic, pyramidal or pyritohedral forms. In some cases the different crystal forms may be intermixed in the same mineral deposits.

Of considerable significance in the development of crystal faces is the *law of the constancy of interfacial angles* in the same mineral species. Wherever a mineral species occurs, whether natural or man-made in the laboratory, the angle of intersection between like faces on the crystals is essentially constant. This is a direct expression of the internal atomic arrangement, and is an important diagnostic feature of the species. This law applies irrespective of the size of the crystal formed. A

10

mechanical goniometer, a semicircular protractor with a movable arm, is used to measure these angles. The reflecting goniometer, a more sophisticated instrument, measures the angles through the use of reflected light from the crystal faces.

Symmetry

Using the crystal axes and the external form of the mineral a description of the symmetry of the crystal form is possible. This is another useful feature in mineral description. Symmetry refers to the repetition of like faces when the crystal is rotated about an axis or is referred to a plane, or a center of symmetry. Such symmetry types are the basis of the division of crystals into systems and classes within the kingdom. Some classes have all symmetry types present, one class lacks the center, axis and plane of symmetry.

Planes of Symmetry. For the ideal geometrical symmetry one half of the crystal is the exact mirror image of the other half. Another descriptive characteristic would be to have a duplicating point at equal distances on the other side of the plane of symmetry when measured normal to it. In the cubic crystal system nine planes of symmetry are possible: three are parallel to the cubic face, and six planes join the opposite cubic edges.

Axes of Symmetry. If a crystal can be rotated about an axis through a certain number of degrees with the result that it repeats the same position in space, this becomes an axis of symmetry. Four different kinds of axes of symmetry are possible. These are in accord with the number of times which the crystal repeats itself in space during a complete revolution of 360 degrees. Note that all axes of symmetry are at right angles to a crystal face and parallel the edge of intersection of two crystal faces. Examples include diagonal or twofold symmetry when the revolution of the crystal is 180 degrees to cause repetition. Trigonal or threefold symmetry results when a revolution of 120 degrees is required for the above effect. Tetragonal or fourfold symmetry is based on a 90 degree revolution, or four repeats in a complete revolution of the crystal. Finally, hexagonal or sixfold symmetry is present when the revolution for repetition is 60 degrees.

Center of Symmetry. Most crystals have a center of symmetry as well as planes and axes of symmetry. Others may have neither of the above, but possess a center of symmetry. This indicates that every face, edge, and solid angle is repeated on the opposite half of the crystal. You can visualize this by turning the crystal about the axis for 60 or 180 degrees and seeing a simultaneous reflection of the faces over a plane normal to the axis of rotation.

Classes of Symmetry. Theoretically, there are thirty-two types of symmetry possible. Among these thirty-two, seven include the largest number of crystallized minerals. Another fourteen are represented distinctly, but several of these are of very rare occurrence. The balance are very rare, if at all present, in nature, but are known in the crystalline salts made in the laboratory.

 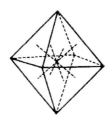 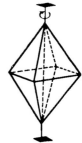

a. Plane of Symmetry   b. Center of Symmetry   c. Axis of Symmetry
                                                       Four-fold

Figure 9.   Examples of symmetry

Twin Crystals

Two or more crystals may be intergrown and certain parts of the different crystals are parallel and other parts of the individuals are in reverse position in relationship to each other. These relationships relate to a deducible law of growth and the final crystal product is a twin crystal. Twinning may occur over a plane common to both and be called a twin plane. A point may have two individual faces symmetrical to it and the twin is then called a twin center. Twin laws indicate whether the twinning is by center, axis or plane and give the crystallographic orientation for either the plane or axis.

Twin crystals may take the form of contact twins or penetration twins. Contact twins have a definite composition surface which separates the two individuals. The twin law is defined by the twin plane. Albite feldspar (plagioclase) beautifully shows this type of twinning, and the name albite twinning is used (figure 10a). Interpenetrating crystals have an irregular composition surface (figure 10c). The twin axis (line) defines the twin law. Multiple twins and repeated twins, of three or more parts, are all twinned according to the same law. Parallel successive composition surfaces form polysynthetic twins. If the successive composition planes are not parallel, cyclic twins result. Sometimes the individuals in polysynthetic twinning are so closely spaced that when a crystal face or cleavage plane crosses the composition planes, distinct striations show because of the reversed positions of adjacent crystals. Additional information on these important twinning laws may be obtained from good mineralogical text books.

Table 1. Most common twins applicable to the different crystal systems.

Crystal System	Twin Planes	Penetration Twins	Twin Axis
Isometric	X	X	X
Tetragonal	X		
Hexagonal (Rhombohedral)	X	X	X
Orthorhombic	X		
Monoclinic	X	X	X
Triclinic	X		X

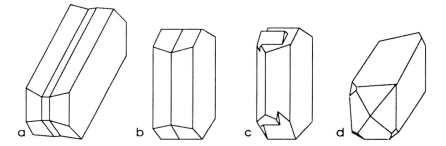

Figure 10. Some twinning types: a) albite type twinning, b) contact twin, c) interpenetration twin, d) saveno type twin

CRYSTAL CHEMISTRY

As an introduction to this section, a brief description is given of the chemical elements, their symbolization and relative abundance in the earth's crust. Table 2 lists the various elements of particular interest to us. While over 100 elements are known, only those pertinent to mineralogical study are included. The table has three major divisions. Part A is an alphabetical listing of the elements in nature, with their corresponding commonly used chemical symbols. This will assist you in the use of this useful alphabet of science.

Part B of the table includes an alphabetical listing of the chemical symbols with their corresponding chemical name. Thus, if you read a symbol and do not know its meaning, you may turn to this table and identify it there. Note that subscripts in the chemical formula refer to the number of atoms of the elements present in a unit cell of the mineral. For example calcite, a common mineral, has the composition of $CaCO_3$, or one atom of Ca (calcium), one of carbon (C), and three of oxygen (O). All minerals have fairly constant composition within certain limits. With a little practice you will learn the composition of the minerals you identify. These compositions are always included in the mineral descriptions of the handbook.

Part C of the table is a listing of the same elements as in the two parts above, tabulated by symbol and decreasing in abundance in the earth's crust. The most abundant minerals will obviously contain silica, oxygen, iron, calcium, sodium, potassium, iron and some titanium. This is confirmed by Table 14 which lists the most abundant minerals of igneous rocks and their estimated percentages. Each of these minerals is included in the mineral charts and in the descriptive section of the handbook together with many other less abundant, but very interesting mineral species.

Crystal Chemistry – An Overview

We are indebted to Dana's *Manual of Mineralogy* and its lucid description of crystal chemistry and mineralogy for parts of this chapter. The reader is referred to this excellent text for more details. We have known for many years that definite crystal morphology, or external form, and chemical composition are related. It was only after x-ray diffraction methods became available to the mineralogist that the importance of these factors was solidly recognized. The new science of crystal chemistry has since developed and has clarified the relationships between chemical composition, internal atomic structure, and physical properties in crystalline materials such as minerals. The end result has been one of simplification and clarification of the science of mineralogy. Crystal chemistry studies may ultimately lead to the integration and understanding of many presently unrelated facts of descriptive mineralogy.

All minerals are composed of atoms from those consisting of single atomic composition such as gold to those of complex composition involving several different types of atoms. The latter, based on electrical charge, are divided into ions, with cations, positively charged ions, and anions, negatively charged ions.

In the Berzelian classification of minerals, chemical composition is the main basis. Minerals are placed in chemically related classes depending on the dominant anion (negative charged ion) or anionic group, such as the combined C (carbon) plus 3 O (oxygen) which is the basis of the carbonate group.

A person desiring to know more about minerals should understand the chemical components, or elements, of which they are composed. They should also know something about atoms, of which elements are made, and their general structure, since an atom is the smallest subdivision of matter which retains elemental characteristics. Yet, even the atom is subdivided into smaller components such as a very small nucleus made up of protons and neutrons surrounded by electrons spread through a much larger space. Let us examine these various components of matter and see how they determine the kind and character of the minerals we study.

Table 2. Elements Of The Earth's Crust (A, B, C)

A. Elements of the earth's crust alphabetized by name and with symbols shown.

Aluminun	Al	Holmium	Ho	Ruthenium	Ru
Antimony	Sb	Hydrogen	H	Samarium	Sm
Argon	Ar	Indium	In	Scandium	Sc
Arsenic	As	Iodine	I	Selenium	Se
Barium	Ba	Iridium	Ir	Silicon	Si
Beryllium	Be	Iron	Fe	Silver	Ag
Bismuth	Bi	Lanthanum	La	Sodium	Na
Boron	B	Lead	Pb	Strontium	Sr
Bromine	Br	Lithium	Li	Sulphur	S
Cadmium	Cd	Lutetium	Lu	Tantalum	Ta
Calcium	Ca	Magnesium	Mg	Tellurium	Te
Carbon	C	Manganese	Mn	Terbium	Tb
Cerium	Ce	Mercury	Hg	Thallium	Tl
Cesium	Cs	Molybdenum	Mo	Thorium	Th
Chlorine	Cl	Neodynium	Nd	Thulium	Tm
Chromium	Cr	Nickel	Ni	Tin	Sn
Cobalt	Co	Niobium	Nb	Titanium	Ti
Copper	Cu	Nitrogen	N	Tungsten	W
Dysprosium	Dy	Osmium	Os	Uranium	U
Erbium	Er	Oxygen	O	Vanadium	V
Europium	Eu	Palladium	Pd	Ytterbium	Yb
Fluorine	F	Phosphorous	P	Yttrium	Y
Gadolinium	Gd	Platinum	Pt	Zinc	Zn
Gallium	Ga	Potassium	K	Zirconium	Zr
Germanium	Ge	Praseodymium	Pr		
Gold	Au	Rhenium	Re		
Hafnium	Hf	Rhodium	Rh		
Helium	He	Rubidium	Rb		

B. Elements of the earth's crust alphabetized by symbol, name, valence (V), and ionic radius (IR), latter for 6-fold coordination.

		V	IR			V	IR			V	IR
Ar	Argon			H	Hydrogen	+1	2.08	Rb	Rubidium	+1	1.57
Ag	Silver	+1	1.23	He	Helium			Re	Rhenium	+4	0.71
Al	Aluminum	+3	0.61	Hf	Hafnium	+4	0.79	Rh	Rhodium	+2	0.86
As	Arsenic	+5	0.58	Hg	Mercury	+2	1.10	Ru	Ruthenium	+3	0.69
Au	Gold	+1	1.37	Ho	Holmium	+3	0.97	S	Sulphur	-2	0.20
B	Boron	+3	0.20	I	Iodine	-1	2.13	Sb	Antimony	+5	0.69
Ba	Barium	+2	1.44	In	Indium	+3	0.88	Sc	Scandium	+3	0.83
Be	Beryllium	+2	0.35	Ir	Iridium	+4	0.66	Se	Selenium	+2	1.88
Bi	Bismuth	+3	1.10	K	Potassium	+1	1.46	Si	Silicon	+4	0.34
Br	Bromine	-1	1.88	La	Lanthanum	+3	1.13	Sm	Samarium	+3	1.04
C	Carbon	+4	0.15	Li	Lithium	+1	0.82	Sn	Tin	+4	0.77
Ca	Calcium	+2	1.08	Lu	Lutetium	+3	0.93	Sr	Strontium	+2	1.21
Cd	Cadmium	+2	1.03	Mg	Magnesium	+2	0.80	Ta	Tantalum	+5	0.72
Ce	Cerium	+3	1.09	Mn	Manganese	+2	0.90	Tb	Terbium	+3	1.00
Cl	Chlorine	-1	1.72	Mo	Molybdenum	+4	0.73	Te	Tellurium	-2	2.21
Co	Cobalt	+2	0.83	N	Nitrogen	+5	0.13	Th	Thorium	+4	1.08
Cr	Chromium	+3	0.70	Na	Sodium	+1	1.10	Ti	Titanium	+4	0.69
Cs	Cesium	+1	1.78	Nb	Niobium	+5	0.72	Tl	Thallium	+1	1.58
Cu	Copper	+1,+2	0.9, 0.8	Nd	Neodynium	+3	1.08	Tm	Thulium	+3	0.95
Dy	Dysprosium	+3	0.99	Ni	Nickel	+2	0.77	U	Uranium	+4,+6	1.08,0.3
Er	Erbium	+3	0.96	O	Oxygen	-2	1.32	V	Vanadium	+3	0.72
Eu	Europium	+2	1.12	Os	Osmium	+4	0.69	W	Tungsten	+4	0.68
F	Fluorine	-1	1.25	P	Phosphorous	+5	0.25	Y	Ytrrium	+3	0.98
Fe	Iron	+2,+3	.86,0.74	Pb	Lead	+2	1.26	Yb	Ytterbium	+2	1.13
Ga	Gallium	+3	0.70	Pd	Palladium	+2	0.94	Zn	Zinc	+2	0.53
Gd	Gadolinium	+3	1.02	Pr	Praseodymium	+3	1.09	Zr	Zirconium	+4	0.80
Ge	Germanium	+4	0.62	Pt	Platinum	+2	0.96				

C. Abundance of elements in the earth's crust in decreasing ppm (parts per million) or grams/metric ton.

No.	El.	ppm	No.	El.	ppm	No.	El.	ppm
1	O	445 000	27	Nd	40	53	Ge	1.5
2	Si	272 000	28	La	35	54	Ho	1.3
3	Al	83 000	29	Y	31	55	Mo	1.2
4	Fe	62 000	30	Co	29	56	W	1.2
5	Ca	46 600	31	Sc	25	57	Tb	1.2
6	Mg	27 640	32	Nb	20	58	Tl	0.7
7	Na	22 700	33	N	19	59	Tm	0.5
8	K	18 400	34	Ga	19	60	I	0.46
9	Ti	6320	35	Li	18	61	In	0.24
10	H	1520	36	Pb	13	62	Sb	0.2
11	P	1120	37	Pr	9.1	63	Cd	0.16
12	Mn	1060	38	B	9.0	64	Ag	0.08
13	F	544	39	Th	8.1	65	Hg	0.08
14	Ba	390	40	Sm	7.0	66	Se	0.05
15	Sr	384	41	Gd	6.1	67	Pd	0.015
16	S	340	42	Er	3.5	68	Pt	0.01
17	C	180	43	Yb	3.1	69	Bi	0.008
18	Zr	162	44	Hf	2.8	70	Os	0.005
19	V	136	45	Cs	2.6	71	Au	0.004
20	Cl	127	46	Br	2.5	72	Ir	0.001
21	Cr	122	47	U	2.3	73	Te	0.001
22	Ni	99	48	Sn	2.1	74	Re	0.0007
23	Rb	78	49	Eu	2.1	75	Ru	0.0001
24	Zn	76	50	Be	2.0	77	Rh	0.0001
25	Cu	68	51	As	1.8			
26	Ce	66	52	Ta	1.7			

IR = Ionic radius in Å (angstrom units)
El. = Element

Almost 200 years ago Dalton, an English chemist, after intensive research, proposed that all elements consist of very small particles called atoms. He further stated that for a given elemental substance the atoms are alike and exhibit identical chemical properties. Different elements have different chemical properties, and in an ordinary chemical reaction, no atom disappears or changes into another element.

A compound is a substance made of two or more elements combined in definite and constant amounts expressed as whole numbers or simple fractions. In water, for example, there are always two atoms of hydrogen (H) to one of oxygen (O) supporting the law of constant composition. Dalton also concluded that when two elements combine to form two different compounds, the weights of one which combine with a fixed weight of another is always in a simple ratio of whole numbers, e.g., 1:2, 3:2, etc. This is the law of multiple proportions.

Research by other scientists, about the turn of the century, revealed the atom actually consists of at least two major units: 1. electrons and, 2. a nucleus. Electrons are present in all atoms and make up only about 1/2000 the mass of the atom and its size is only 1/10,000 that of the atom. The electron carries an electrically negative charge of the same amount as the nucleus. All atoms show an integral number of electrons, ranging from one to over 100, depending on the particular element. The electrons occur in the outer parts of the atom.

The geometric arrangement of an atom in a simple model is somewhat similar, disregarding scale, to the planets revolving about the sun. The sun would be analogous to the nucleus, and the planets to the electrons. In terms of scale, the atoms are extremely small, only 0.46 angstrom units diameter, (1/375,000,000 of an inch) for the hydrogen ion which is the smallest. It consists of a proton, a positive particle in the nucleus, and one electron moving about the nucleus. Even the largest cation, cesium, has a radius of only 2.62 angstrom units. The proton carries the positive unit charge of electricity, the electron has the negative charge, and the neutrons, also a part of the nucleus, are electrically neutral. Because an atom is chemically neutral, there must be an equal number of protons and electrons in the atom to maintain electrical neutrality.

Even though solids made of such atoms appear very dense, it is apparent that at the atomic scale, most of the solid mineral is in reality open space between the nucleus and electrons of the individual atoms.

If all elements are made up of the same primary building blocks of neutrons, protons, and electrons, what makes them so different? The difference must relate to the total number of electrons of the atom and/or corresponding number of protons.

Hydrogen has one electron of a negative charge and a simple positive charge. Helium, next higher in atomic weight, has two electrons and an equal positive nuclear charge. In recent years research has shown that nuclei of the hydrogen element may differ in mass, that is extra mass is present in the nucleus., but their positive and negative charges stay unchanged. Hydrogen, for example, has two varieties: one with a single mass and another with double mass. Many elements show such differences in mass and we call these isotopes – with similar chemical properties, but different number of units of mass in the nucleus. Tin, as an example, has 10 isotopes and carbon three.

Only a few elementary substances are made up of individual atoms and also have the basic structured unit of which the substance is composed. The noble gases, which are inert and non-reactive to other elements, are such. These include helium (He), neon (Ne), argon (Ar), krypton (Kr), xenon (Xe), and radon (Rn). All other elementary and compound substances consist of two other structured units derived from atoms.

The molecule is the fundamental structural unit of most volatile or gaseous substances. This is an aggregate of atoms held together by strong forces known as chemical bonds. The force between individual molecules is relatively weak, hence there is usually little interaction with each other. Examples in gases include oxygen (O), which is a diatomic (O_2) or two atoms combined; carbon-dioxide, with one carbon (C) and two oxygens (O_2) to form the molecule CO_2. Finally, water has two hydrogen (H_2) atoms and one oxygen atom (O) and forms H_2O.

To understand bonding between various elements to form simple or complex compounds, such as minerals, let us recall our understanding about attraction of electric charges: Like charges repel, opposite charges attract (figure 11).

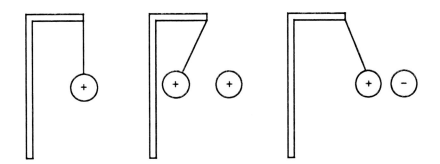

Figure 11. Reaction of electrically charged bodies.

Armed with this knowledge, we are now able to discuss the principle of chemical bonding.

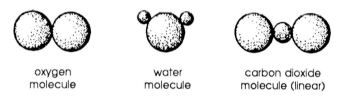

Figure 12. Schematic drawing of oxygen, water, and carbon dioxide molecules.

An electron may be released from an electrically neutral atom and leave behind a positively charged particle somewhat smaller than the original atom and called a cation. If, on the other hand, an electron is added to certain neutral atoms, it becomes negatively charged and is known as an anion. Either positively or negatively charged atoms take the term ion. Examples of cations include sodium (Na^{1+}), calcium (Ca^{2+}) and hydrogen (H^{1+}). Familiar anions include chlorine (Cl^{1-}), fluorine (F^{1-}) and others.

Many compounds consist of positively and negatively charged ions making up their basic structural units. Possibly most familiar is sodium chloride (NaCl).

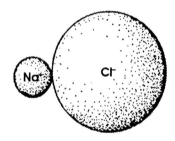

Here ions of sodium (Na^{1+}) and chlorine (Cl^{1-}) are combined as shown below. Their structure is constant throughout the sodium chloride mineral, halite.

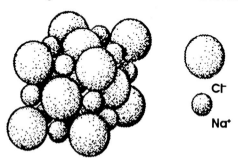

Fluorite, a calcium fluoride (CaF_2), is also a salt. The basic structure is cubic, with fluorine (F^{1-}) on the corners of the cubes and calcium (Ca^{2+}) at the center of alternate cubes. Each of these minerals, halite and fluoride, are ionic substances which are held together by a three dimensional network of chemical bonds holding the appropriately charged ions together. These electrostatic attractions are very strong.

The brief discussion above is a prologue to how elements are bound together to form minerals. From the foregoing we recognize the importance of electrical charges among the elements. So it is in mineral formation. The intensity and type of these electrical charges determine many of the physical and chemical properties of minerals. If the electrical bond is strong, the crystal is hard, the boiling point is higher, and thermal expansivity is less because of the strong electrical bonds. Such factors help to explain the differences in properties between graphite (C) and diamond (C).

The electrical forces, or chemical bonds, belong to one of four principal bond types: ionic, covalent, metallic and van der Waals', with transitions between all types. In mineral genesis nature acts to form the best geometry of structural units in space combined with electrical neutrality and minimum lattice energy. These arrangements determine the physical and chemical properties of the mineral.

The Ionic Bond – All atoms tend to achieve a stable configuration, which is to have all possible electron sites filled. As noted, helium, neon, and argon have achieved this state. Krypton and xenon also have their electron sites filled and are essentially inert gases. Other elements have one or more additional electrons and form a stable configuration with a negative charge. Chlorine (Cl^{1-}), bromine (Br^{1-}), and others are examples. Some elements form divalent anions with a double negative charge, such as oxygen (O^{2-}), sulphur (S^{2-}) and others.

If sodium ions (Na^{1+}) and chlorine (Cl^{1-}) ions are mixed in a solution, continuous collision keeps them separate until the temperature is lowered or solvent volume decreased and the attraction of opposite electrical charges of the ions exceeds the collision disruptive forces. When a large enough number of the cations and anions attach themselves to a growing nucleus, crystal growth occurs. In this nucleating mass is the embryonic mineral with all of its characteristic features. This type bonding is defined as the ionic bond. It is characteristic of ions which dissolve in polar solvents, such as water and form conducting solutions with free ions.

Some properties of ionic-bonded crystals are moderate hardness, moderate specific gravity, fairly high melting and boiling points, and poor conductors of heat and electricity. Because of the uniformity of the electrostatic charge, the resultant crystal has high symmetry.

We may apply some generalizations to cations and anions. For example, the size of the ion depends on the state of ionization. Cations, generally, are smaller, more rigid, and less expansible than anions. An example is the marked contrast between sulfur (S^{2-}) at 1.85 Å and (S^{6+}) at 0.30 Å. Anions are larger, more easily polarized, and less likely to have variation in size with change in valence, or charge, than cations.

The bond strength is the amount of energy required to break the bond and is dependent on center-to-center spacing of the ions and their total charge. If this spacing or the interionic distance is small, e.g., NaCl is 2.81 Å versus a NaI (Sodium Iodide) at 3.23 Å, the melting point decreases from 801° to 651° C, respectively, indicating an inverse relationship. Likewise, hardness of a mineral is dependent on interionic distance (ID). BeO (beryllium oxide) with an ID of 1.65 Å has a hardness of 9.0 on the Mohs scale, yet BaO (barium oxide) with 2.77 Å has a hardness of only 3.3 Mohs hardness.

Covalent Bond – Electron sharing ibetween elements develops the strongest of the chemical bonds. Chlorine, as an example, is a very chemically reactive anion and seizes the nearest ion, usually another chlorine ion. The two ions interact and the same resultant stable configuration develops as in an inert gas. Similar electron sharing in atoms within minerals results in general insolubility, a greater stability, very high melting and boiling points and non-conducting of electricity. The local and sharply directional bonds formed result in a lower symmetry for formed crystals than when ionic bonding occurs.

Some elements, especially those toward the middle of the periodic table such as carbon (C), silicon (Si), aluminum (Al) and sulfur (S), have two, three and four vacant electron spaces. Not all of the bonds are used in linking just to a neighboring atom. These elements tend to link together in covalent bonding with adjacent atoms and form fixed forms of set dimension. These in turn link together and form aggregates or groups.

We earlier noted the contrast between the polymorphs diamond and graphite. In diamond the carbon atom has four electon vacancies. These are shared with four other carbon cations and form a very stable, firmly bonded tetrahedral form with carbon atoms at the four apices, or pointed ends. The bond is very strong in the vicinity of the shared electrons and results in a very rigid and polarized structure, forming the hardest known natural substance. Yet, in the mineral graphite, with exactly the same chemical composition, cohesion between carbon cations through strong covalent bonding is limited within a sheetlike structure, but the van der Waals' bond, a much weaker force, keeps the sheets together. Hence, there is an ease of excellent cleavage, or parting, along this weakly bonded sheet interface and the mineral is relatively soft and of greasy feel. Such examples indicate that minerals have more than one bonding force. As a general rule the alkali halides and alkali oxides tend to have ionic bonding, while like-atoms or atoms close together in the periodic table will be covalent.

van der Waals' Bond – The very weak bond which ties neutral molecules (one or more combined atoms) and neutral structural units together into a lattice through small residual charges on their surfaces is the van der Waals' bond. It is most uncommon in minerals and, if present, such as in graphite and mica, defines a zone of ready cleavage and low hardness.

Metallic Bond – When atomic nuclei are bound together by the aggregate electrical charges of the electrons that surround them, this is the metallic bond. The electrons are so weakly tied to the metal structure that they may drift through it, or even out of it without disrupting the structure. Light energy may actually free a large number of electron from certain metals, the so-called photoelectric effect commercially used in light meters, door openers, burglar alarms and other such devices. The close metallic bonding results in high plasticity, ductility, tenacity, electrical conductivity, and low hardness. Melting and boiling points also relate directly to this bonding. Gold, silver, copper and a few other native metallic minerals have metallic bonding.

Coordination Principle and Radius Ratio – The ions of the different elements vary in size of their ionic radii (Table 2). The positively charged cations are usually much smaller than the negatively charged anions. Given opposite charged ions which unite to form a crystal structure, an ion tends to gather to itself as many ions of the opposite sign as the ion size permits. The gathered anions cluster about the central coordinating cation so that their centers are at the apices, or pointed ends, of a regular polyhedron. The coordination number (CN) is the number of anions in the polyhedron. For the mineral fluorite (CaF_2) each calcium (Ca^{2+}) ion is at the center of eight fluorine ions and the CN is 8 in respect to fluorine. Yet, in the same mineral each fluorine ion has only four close calcium neighbors and is in 4-fold coordination (CN-4) in respect to calcium.

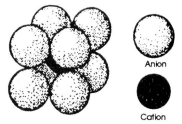

Anion

Cation

Figure 13. 8-fold (cubic coordination) of anions about a cation.

Strongest forces exist between ions which are nearest to each other. The geometric form and coordination number depend on the relative sizes of the coordinated ions. The relative sizes are expressed as radius ratios. As an example, for CaF_2 the ionic radius for Ca is 0.99 Å and for F (fluorine) 1.36 Å, hence $RCa/RF = 0.99/1.36 = 0.73$. Ratio determinations may be calculated for other cation anion pairs or combinations. When these are compared we find that there are limiting coordination conditions for various ratios as shown below.

Table 3. Coordination Numbers (CN) and Radius Ratios

Coordination	Maximum Stability Radius Ratio	CN	Example
Midpoint Cubic Coordination	1.000	12	Cesium & Potassium Ions
Cubic Coordination	0.732 - 1.000	8	Fluorite
Octahedral Coordination	0.414 - 0.732	6	Na, Cl in halite Ca and CO_3 in calcite
Tetrahedral Coordination	0.225 - 0.414	4	SiO_4 groups in silicates, ZnS (sphalerite) structure
Triangular Coordination	0.155 - 0.225	3	Carbonate (CO_3), NO3 and BO_3 ionic groups
Two-fold	Very rare in ionic bonded crystals Uranyl and cuprite (Cu_2O) groups	2	Rare in ionic bonded

We end this discussion with a comment on the variation in composition of individual minerals, which in turn relates to some of the above discussion. Substitution of an ion or ionic group of a different element may occur in a given structure. This is called ionic substitution and the extent to which it occurs depends on several factors. Most important is ionic size. For example, ions of two elements can substitute for each other if the ionic radii are similar or differ by less than 15%. If more than this, substitution is rare. Temperature of mineral formation may also affect substitution within the above limits. The higher the temperature, the greater the thermal activity and the less rigorous the space requirements within the lattice structure of the mineral. Crystals formed at high temperatures may show extensive ionic substitution, which is not possible at lower temperatures. A similar valence or electrical charge favors substitution, but is not a requirement as long as electrical neutrality can be maintained by other ions.

Complete ionic substitution may take place in mineral groups of exactly equal structure (isostructural). For example siderite, $FeCO_3$, can have Mg^{2+} enter the lattice structure of the mineral in any proportion, or Fe^{2+} may enter the lattice of magnesite, $MgCO_3$, in all proportions. The valence charge is the same. The most abundant mineral group in the world, the plagioclase feldspars, have a type of solid solution or substitution of Na and Ca from the calcic rich anorthite to the sodic rich albite, i.e., $NaAlSi_3O_8 \rightarrow CaAl_2Si_2O_8$.

Silicate Structures

The individual type(s) of interatomic linkages or chemical bonds in minerals, metallic, ionic or polar, covalent and van der Waals, imparts characteristic properties to the matter in which it occurs. More than one type of chemical bond may be present

In a single compound: 1. mixed and 2. like bonds. Several physical properties are the result of the mixed type such as hardness, melting point, and strength. Each physical characteristic is determined by the weakest bonds. These are directly affected by increased mechanical or thermal stress. Still, the classification of bonding is arbitrary and many compounds may have intermediate types. For example, the silicates are neither purely ionic nor covalent but intermediate. As earlier noted, the geometry assumed by any solid is such that the entire crystal lattice tends to form with minimum potential energy.

Almost all common minerals have ionic structures, with the important and large oxygen anions and all other elements, except the halogens, as cations. You will note that the oxygen anions are so large compared to most cations, that the structure of a mineral is mainly the packing of oxygen ions with cations in the interstices. Recall that the ion tends to surround itself with ions of opposite charge, and the number grouped about the central ion depends upon the radius ratio between the two (Table 3). Because the radius ratio of some cations lies near the theoretical border between two types of coordination, they may occur in both. Aluminum is a case in point. Finally, electrical neutrality and geometrical stability are prime factors in determining the structure of ionic compounds.

In the case of silicates, which are the most abundant of all mineral types, silicon lies between four oxygen atoms. This structural block is universal in the silicates. The four oxygens always occur at the corners of a tetrahedron of almost constant size and regular shape (figure 14).

The different types of silicates result from the various ways in which the silicon-oxygen tetrahedra are linked to each other. These various linkages, or lack of, form several silicate structural types.

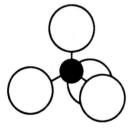
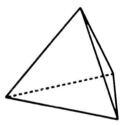

Figure 14. The silicate tetrahedron represented in two different ways.

Independent tetrahedral groups, or *nesosilicates,* have the composition SiO_4. The mineral olivine, variety forsterite (Mg_2SiO_4), is typical of this group. Other common nesosilicates are the Al_2SiO_5 group of andalusites, sillimanite and kyanite, staurolite, topaz, the garnet groups, zircon and sphene (figure 15).

The *sorosilicates* are formed by two tetrahedra sharing a common oxygen between the ions. The composition is Si_2O_7 for this subclass. Common minerals include idocrase, epidote, and hemimorphite. If more than two tetrahedra are linked and a closed or ring-like structure is formed, the composition will be Si_nO_{3n}. Three-linked tetrahedra include benitoite, $BaTiSi_3O_9$, and beryl with six-linked tetrahedra, $BeAl_2Si_6O_{18}$.

You will recognize the possibility of the development of chain-like structures of two types and of indefinite extent: a) single chains, and b) double chains. These are called *inosilicates.* In the single chain the Si:O ratio = 1:3. This type structure characterizes the large pyroxene mineral group. If alternate tetrahedra in two parallel single chains are linked, the Si:O ratio is 4:11. This structure is characteristic of the well-known, and quite abundant amphibole group. (figure 15).

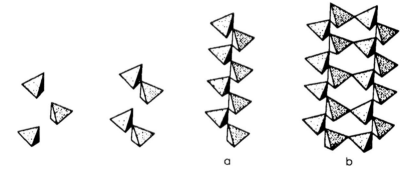

Figure 15. Nesosilicate, Sorosilicate and Inosilicate structures: a) single chain, b) double chain. (Drawings modified to 3D from Simons and Schusters *Guide to Rocks and Minerals.*)

A sheet structure, known as the *phyllosilicates*, develops when three oxygen ions of each tetrahedra are shared with adjacent tetrahedra. This is the double-chain inosilicate structure, extended indefinitely in two directions forming a sheetlike structure. In more detail the sheets have a hexagonal planar framework which results in a false hexagonal habit. Perfect basal cleavage occurs parallel to the plane of the sheet, where the sheets are interconnected by weak van der Waals bonds. Such linkage yields a ratio of Si:O of 2:5. All micas and clay structures are built on this basic unit, as well as talc.

The tectosilicates are a three dimensional or polymerized network with a SiO_4 tetrahedron sharing all its corners with other tetrahedra. The Si:O ratio is 1:2. Quartz and other silica forms such as tridymite and cristobalite have this structure. Because Al^{3+} may substitute for Si^{4+}, additional positive ions are required to maintain electrical neutrality. The very abundant feldspars, nepheline, and less abundant zeolites, are examples.

Figure 16. Tectosilicate structures: quartz and sodic feldspar

Because of the extreme importance of the SiO₄ tetrahedral linkages, or its type of polymerization during magma crystallization, we present a reaction series diagram which may help to clarify the relationships of chemistry and geometry in the development of the very important silicate minerals from igneous mets.

Bowen and his co-workers reported on productive research results on the characteristics of magmatic crystallization. They were able to show that the common minerals of igneous rocks, which are the most abundant in the mineral kingdom, can be arranged into two series: 1: a discontinuous reaction series made up of the ferromagnesian minerals, or minerals containing iron and magnesium, and 2: a continuous reaction series of the feldspars. Figure 17 illustrates this reaction series.

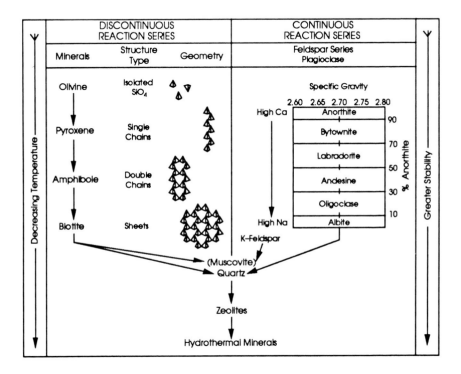

Figure 17. Bowen's reaction series (modified)

For the discontinuous series the crystallization begins at the highest temperature with olivine, one of the isolated SiO₄ tetrahedral groups, or nesosilicates. The temperature decreases and if still molten magma remains, the olivine completely reacts with the melt and gradually changes to pyroxene, the single chain tetrahedra of SiO₄. At the same time calcic rich feldspar continues to react with the melt and sodium becomes part of the feldspar composition. If some of the olivine and bytownite are removed from the crystallizing system, e.g., by settling of the olivine in the magma, or the floating of the less dense feldspar phenocrysts to the top of the magma chamber, then the reaction series continues to take place in the molten magma. Amphiboles and labradorite feldspar may crystallize. If complete crystallization takes place at this stage, a rock with the composition of amphibole-labradorite forms. The latter rock is an andesite or diorite depending on its cooling rate and the texture developed.

The longer the magma remains fluid, the more the reaction series proceeds. Granites and pegmatites, zeolitic rich rocks, and finally, hydrothermal minerals may form, including the metallic ore minerals. Each of these is derived from the still molten mass or the residual fluids of the original magma. Pegmatites probably form as a result of residual high water content and other mineralizer fluids in the late stage magma. These permit lower temperatures of crystallization, high mobility of the various ions or elements, and endow a greater fluidity to the end-stage magma to form the unique pegmatites and the ore minerals.

From olivine to quartz, all structural forms develop and may disappear. You will note the geometric progression from the isolated tetrahedra of SiO_4 first, then the single chains, double chains, sheets, and finally the three-dimensional networks or tectosilicates. This reaction series is a marvel of orderly development and geometric stability for the conditions extant at the time of the crystallization of the magma. Herein lies the secrets of not only structure of silicates minerals, but composition as well. For greater detail on this key structural buildup of nature, detailed writings on the structure of minerals would be appropriate.

PHYSICAL PROPERTIES OF MINERALS

Seven diagnostic properties of minerals are described. These same properties are included in each of the individual mineral descriptions. The properties include color, cleavage, hardness, luster, streak, specific gravity and tenacity. Tenacity includes sectility, malleability, brittleness, flexibility, and elasticity. Elasticity is a property common to all minerals but is optically visible in only a very few. A single physical property may not permit specific identification, but several such properties in combination will. Other specific properties of minerals such as fusibility, elemental composition and chemical properties are described later in the book.

Color: The color of a mineral depends upon the selective absorption which it exerts upon the light reflected by it. If a mineral is green, for example, it reflects all the light waves except those which combined yield the sensation of green. Color is an important diagnostic property of each mineral. In studying the identification charts you will recognize that some individual minerals appear on several color charts. Such color variations may be present in the same outcrop, or in rare occasions in the same mineral specimen.

Cleavage: Cleavage is the second major primary identifier on the mineral identification charts. This physical property is the tendency of a mineral to break or cleave along crystallographic planes. Thus, cleavage is a more or less smooth surface of breaking in the mineral. The cleavage planes are reflections of possible crystal faces and, in turn, planes of the internal atomic structure of the mineral. These cleavage planes show minimum values of cohesion parallel to the cleavage surface.

Cleavage is readily recognized by the mirror-like reflection of light from the mineral surface when it is rotated in strong light. Keep in mind that cleavage is a broken surface of planar form in contrast to a primary crystal face.

The mineral identification charts consider only the presence of distinct or indistinct cleavage in the mineral. Remember that several cleavage types are possible in various minerals. Some show one-direction cleavage, others two-direction at right angles, two direction not at right angles, and even three direction cubic, three direction rhombohedral, and others. Examples of some of the more common cleavages are shown in figures 18.

Each individual mineral description includes a comment on the character of, or lack of cleavage, for that specific mineral, and the kind of cleavage to expect. For your practical use, however, the presence or absence of cleavage is generally sufficient for most mineral identifications.

Luster: In the identification tables we have used METALLIC and NON-METALLIC properties of the mineral as a major identifier. The metallic luster is a reflectance characteristic of metals such as copper, iron, tin, and others. The non-metallic luster includes several varieties such as adamantine, or that of the highly refractive index such as for diamond; vitreous luster, or that of broken glass or of the freshly fractured surface of the mineral quartz; resinous; greasy luster, like that of oiled glass; pearly luster; and silky luster, or a reflectance similar to that of silk. The latter is generally related to a very fibrous mineral.

The explanation for the identification chart shows which minerals have metallic or non-metallic luster. The length of this line or bar is also an indication of the hardness range of each mineral.

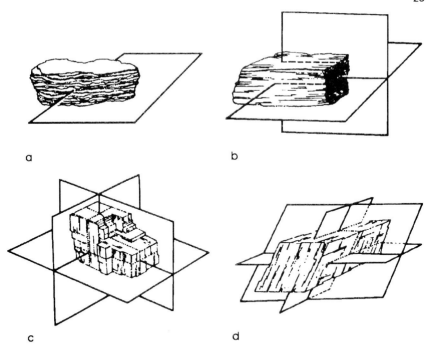

Figure 18. Examples of mineral cleavage: a) one-direction, b) two direction, c) 3-direction at right angles, d) 3-direction not at right angles.

Hardness: The resistance a mineral offers to scratching, or it's hardness, can be determined by scratching it with a material of known hardness, or a mineral of known hardness. For example, a mineral of known hardness will scratch one of lesser hardness. In making the hardness test remember that a scratch cannot be rubbed off. The powder formed in the hardness test may resemble a scratch, but if it can be rubbed away, and no scratch impression remains, the mineral has not been scratched. In this case it would suggest that the material used for the hardness test is softer than the tested or unknown material.

The hardness scale based on the relative hardness of certain minerals is known as the Mohs Scale of Hardness. It is the present standard for classifying the hardness of various minerals. While the scale is not quantitative in the absolute sense, it is most useful in identifying this special physical property of individual minerals. Ten minerals with different and progressively increasing hardness make up the scale. These, in order of increasing hardness, include: 1. Talc, 2. Gypsum, 3. Calcite, 4. Fluorite, 5. Apatite, 6. Feldspar, 7. Quartz, 8. Topaz, 9. Corundum, 10. Diamond. As a supplement to the Mohs Scale of Hardness, the fingernail has a hardness of 2.5, a copper penny 3.0, and a glass or pocket knife blade about 5.5. A good steel file is 6.0 on the scale.

The description of the individual minerals always includes the Mohs hardness number. Some minerals show different hardness numbers for the various cleavage faces. These variations relate to the internal atomic arrangement within the mineral. An example is kyanite, it has a hardness of 5 parallel to its length, and 7 across the width of the crystal. Calcite is another example. It has a hardness of 3 on all surfaces except the base. The latter can be scratched by a fingernail of 2.5 hardness.

Figure 19 shows the Mohs Scale of Hardness and the minerals making up the scale in their position of relative hardness. In addition, a more absolute measure of hardness, derived from instruments designed to obtain quantitative measurements of this physical property, is also shown. This is shown as the Absolute hardness on the right side of the graph.

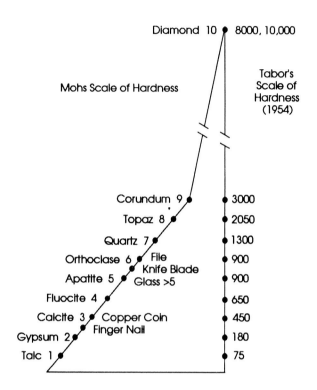

Figure 19. Mohs scale of hardness and Tabor's absolute scale of hardness.

Specific Gravity: The weight of the mineral divided by the weight of an equal volume of water determines the specific gravity of the substance. The general property is shown on the identification charts by a range of specific gravities from 1.0-3.0; 3.0-4.0; 4.0-5.0 and greater than 5.0. A more quantitative determination, or a range, is shown in parentheses behind each mineral name in the chart. These values are also included in the mineral descriptions.

Sophisticated equipment such as the Jolly Balance, or the Berman Micro-density Balance are available for specific gravity determination of unknown minerals. A recently developed field model is also available.

Specific gravity (G) is obtained directly by determining the weight of the mineral in air (w), then obtaining the weight when the mineral is submerged in water (w'). The difference between the two weights (w–w') is, of course, the weight of the volume of water equal to that displaced by the solid. The specific gravity is then determined by the formula:

$$G = w/w-w'$$

Instructions for use of the Berman micro-balance for very small mineral fragments are included with this precision instrument. Normally, a liquid with lesser surface tension than water, such as toluene, is used and appropriate corrections made for the lesser density in the liquid. Fragments the size of a pinhead can be determined to the second, and even third decimal place for specific gravity.

The Jolly Balance (figure 20) is a spiral spring balance with two weighing pans, an upper one in the air, and a lower one immersed in water. Weights of the mineral are established with a relative reading for the spring alone with the upper pan in air and the lower pan submerged in water (n), then placing the mineral on the pan in the air, leaving the lower pan immersed at the same depth in water and determining the scale reading (N_1). The pan in water is raised out of the water, the sample placed on it and then it is again immersed in water with the sample on it to the same level. A scale reading is then obtained (N_2). The specific gravity, G, is thus:

$$G = \frac{N_1 - n}{N_1 - N_2}$$

Example: n = 35   No mineral on pans, spring scale reading only

N_1 = 65   Relative weight of mineral in air

N_2 = 55   Relative weight of mineral in water

$$G = \frac{65 - 35}{65 - 55} = \overline{3.0}$$

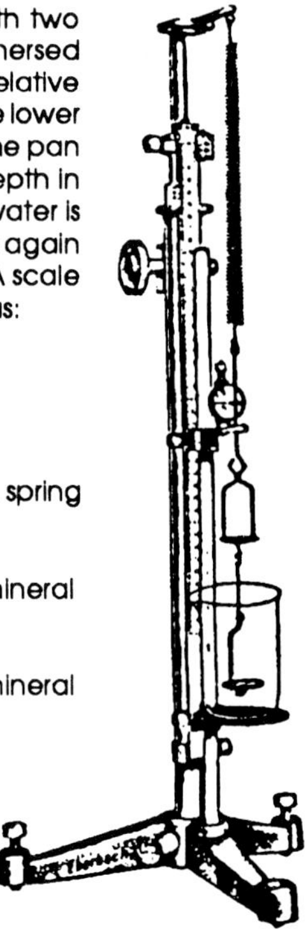

Figure 20.   The Jolly Balance.

Much smaller mineral fragments occur in stream sediments and in rocks. A "sink-float system can be used for those minerals of 2.8 or greater S.G. through the use of heavy liquids. While the dense (4.0-5.0 and very dense > 5.9) cannot be determined singly, they can be singled out as a group which is not large in number. For those minerals with S.G. between 2.89-4.15 the judicious use of the heavy liquids permits fairly precise determinations of this mineral property. The four most commonly used heavy liquids are: 1) Bromoform, $CHBr_3$ (S.G. 2.39); 2) Tetrabromoethane, $C_2H_2Br_4$ (S.G. 2.96); 3) Methylene iodide, $CH2$ (S.G. 3.33); 4) Clerici solution (S.G. 4.15). These heavy liquids are miscible with each other, thus intermediate densities are obtained by diluting one with another.

If a mineral fragment is immersed in a heavy liquid of known density and the mineral floats, the mineral is less dense than the liquid. If it sinks, it is, of course, more dense than the known liquid. In the latter case clean the mineral fragment and try immersion in the next denser liquid. If it falls between the two, the mineral identification charts can be used since specific gravity is the third identifier of five on these charts. If a precise determination of S.G. is required, then intermixing of the heavy liquids in known proportions gives a new liquid of known density in which the mineal can be tested. Practice can bring good results on the determination of this important physical property. Table 4 should also prove very useful in identifying the location within the table of the mineral under investigation.

Table 4 lists the increasing specific gravities of the minerals included in this book.

TABLE 4. SPECIFIC GRAVITY OF MINERALS
(Minerals listed according to increasing Specific Gravity)

1.0 to 2.0 *

Carnellite	(1.6)
Ulexite	(1.65)
Borax-Kernite	(1.7)
Epsomite	(1.7)
Opal	(1.9-2.2)
Sylvite	(2.0)

2.0 to 3.0 *

Chrysocolla	(2.0-2.4)
Bauxite	(2.0-2.6)
Niter	(2.1)
Sulphur	(2.1)
Trona	(2.1)
Chabazite	(2.1-2.2)
Zeolite GP	(2.1-2.2)
Graphite	(2.1-2.3)
Sodalite	(2.1-2.3)
Halite	(2.1-2.5)
Heulandite	(2.18-2.2)
Natrolite	(2.2)
Soda Niter	(2.2)
Analcite	(2.2-2.3)
Serpentine	(2.2-2.6)
Gypsum	(2.3)
Wavellite	(2.3)
Gibbsite	(2.3-2.4)
Apophyllite	(2.3-2.4)
Garnierite	(2.3-2.8)
Brucite	(2.4)
Colemanite	(2.4)
Lazurite	(2.4-2.5)
Leucite	(2.5)
Orthoclase	(2.5-2.6)
Kaolinite	(2.6)
Nepheline	(2.6)
Cordierite	(2.6-2.7)
Vivianite	(2.6-2.7)
Alunite	(2.6-2.8)
Plagioclase	(2.6-2.8)
Turquoise	(2.6-2.8)
Chlorite	(2.6-2.9)
Quartz	(2.65)
Calcite	(2.7)
Glauberite	(2.7-2.8)
Pectolite	(2.7-2.8)
Talc	(2.7-2.8)
Muscovite	(2.7-3.0)
Biotite	(2.7-3.2)
Beryl	(2.75-2.80)
Dolomite	(2.8-2.9)
Wollastonite	(2.8-2.9)
Prehnite	(2.8-3.0)
Boracite	(2.9)
Anhydrite	(2.9-3.0)
Aragonite	(2.9-3.0)
Datolite	(2.9-3.0)
Phenacite	(2.9-3.0)
Amphibole Group	(2.9-3.4)
Hornblende	(2.9-3.4)

3.0 to 4.0 *

Annabergite	(3.0)
Amblygonite	(3.0-3.1)
Diaspore	(3.0-3.1)
Erythrite	(3.0-3.1)
Glaucophane	(3.0-3.1)
Magnesite	(3.0-3.1)
Lazulite	(3.0-3.1)

Actinolite	(3.0-3.2)
Allanite	(3.0-3.2)
Magnesite	(3.0-3.2)
Tourmaline	(3.0-3.2)
Fluorite	(3.0-3.3)
Wad	(3.0-4.3)
Autunite	(3.1)
Andalusite	(3.1-3.2)
Apatite	(3.1-3.2)
Chondrodite	(3.1-3.2)
Spodumene	(3.1-3.2)
Enstatite	(3.1-3.3)
Cummingtonite	(3.1-3.6)
Torbernite	(3.2)
Sillimanite	(3.2-3.3)
Diopside(Augite)	(3.2-3.6)
Pyroxene	(3.2-3.6)
Dumortierite	(3.3)
Idocrase	(3.3-3.4)
Olivine	(3.3-3.4)
Epidote	(3.3-3.5)
Jadeite	(3.3-3.5)
Goethite	(3.3-4.3)
Psilomelane	(3.3-4.7)
Hemimorphite	(3.4-3.5)
Orpiment	(3.4-3.5)
Hypersthene	(3.4-3.5)
Rhodochrosite	(3.4-3.6)
Sphene	(3.4-3.6)
Diamond	(3.5)
Realgar	(3.5)
Topaz	(3.5-3.6)
Chrysoberyl	(3.5-3.8)
Spinel	(3.5-4.1)
Garnet Group	(3.5-4.3)
Carnotite	(3.5-5.0)
Staurolite	(3.6-3.7)
Kyanite	(3.6-3.7)
Rhodonite	(3.6-3.7)
Limonite	(3.6-4.0)
Tyuyamunite	(3.6-4.2)
Strontianite	(3.7)
Atacamite	(3.8)
Azurite	(3.8)
Octahedrite	(3.8-3.9)
Siderite	(3.8-3.9)
Brookite	(3.8-4.1)
Antlerite	(3.9)
Celestite	(3.9-4.0)
Malachite	(3.9-4.0)
Sphalerite	(3.9-4.0)
Corundum	(3.9-4.1)
Willemite	(3.9-4.2)

4.0 to 5.0 *

Chalcopyrite	(4.1-4.3)
Rutile	(4.2-4.3)
Manganite	(4.2-4.4)
Chromite	(4.2-4.6)
Witherite	(4.3)
Smithsonite	(4.3-4.4)
Ilmenite	(4.3-4.5)
Bravoite	(4.3-4.6)
Stannite	(4.4)
Enargite	(4.4-4.5)
Barite	(4.5)
Stibnite	(4.5-4.6)
Covellite	(4.6-4.7)

Pyrrhotite	(4.6-4.7)
Tetrahedrite	(4.6-5.1)
Ilmenite	(4.7)
Zircon	(4.7)
Molybdenite	(4.7-4.8)
Pyrolusite	(4.7-4.9)
Marcasite	(4.9)
Bastnaesite	(4.95)
Greenockite	(4.9-5.0)
Hematite	(4.9-5.3)
Monazite	(4.9-5.3)
Bornite	(4.9-5.4)
Pyrite	(5.0)
Pentlandite	(5.0)

> 5.0 *

Columbite-Tantalite	(5.0-8.0)
Magnetite	(5.2)
Millerite	(5.3-5.7)
Cerargyrite	(5.5-5.6)
Chalcocite	(5.5-5.8)
Pyrargyrite-Proustite	(5.5-5.9)
Jamesonite	(5.5-6.0)
Zincite	(5.6)
Arsenic	(5.7)
Bournonite	(5.7-5.9)
Smaltite-Chloanthite	(5.7-6.8)
Cuprite	(5.8-6.1)
Scheelite	(5.9-6.1)
Arsenopyrite	(5.9-6.2)
Descloizite	(5.9-6.2)
Cobaltite	(6.0-6.3)
Polybasite	(6.0-6.5)
Skutterudite	(6.1-6.9)
Stephanite	(6.2-6.3)
Anglesite	(6.2-6.4)
Bismuthinite	(6.4-6.6)
Tenorite	(6.5)
Cerussite	(6.5-6.6)
Pyromorphite	(6.5-7.1)
Pitchblende	(6.5-8.5)
Vanadinite	(6.7-7.1)
Wulfenite	(6.8)
Cassiterite	(6.8-7.1)
Mimetite	(7.0-7.2)
Niccolite	(7.0-8.0)
Wolframite	(7.1-7.5)
Argentite	(7.3)
Iron	(7.3-7.9)
Galena	(7.6)
Uraninite	(8.0-10.6)
Cinnabar	(8.1)
Copper	(8.8-8.9)
Calaverite-Sylvanite	(9.0-9.4)
Silver	(10.0-11.0)
Platinum	(14.0-19.0)
Gold	(15.6-19.3)

Other Physical Properties of Minerals

If it is not possible to make a specific mineral identification after careful check of the physical properties and reference to the mineral identification charts, then other tests may be necessary. Some of these are excellent for confirming a presumed identified mineral. These additional tests range from relatively simple physical to more complex chemical tests. The descriptive sections follow this order from the simple to the complex.

Magnetic Response: A strong hand magnet is very useful in identifying the four or five minerals which show sufficient magnetic response to be readily recognized. Suspending a small magnet on a thread or string and bringing the mineral close to it is the common magnetic test. The most magnetic mineral is magnetite. Lodestone, the very magnetic variety of magnetite, was once used as a crude compass in ancient times. It is recognized in the field by the broken grains which line up parallel to the magnetic field of the mineral when struck by a hammer.

Pyrrhotite, an iron sulfide mineral, is slightly magnetic and affects a strong hand magnet. Likewise, chromite, a chromic oxide with a molecule of ferrous oxide, also slightly attracts the suspended hand magnet. Small particles of specular hematite are commonly affected by a strong magnet, as also franklinite and even an iron-bearing platinum. The magnetic character of the former two may result from minute magnetite inclusions in them.

The use of a very strong electromagnet permits a distinction of two classes of minerals: diamagnetic and paramagnetic. The latter are attracted by the magnetic field and the former repelled by it. The properties are commonly used in the laboratory and industry to separate various minerals. Quantitative separations of the diamagnetic and paramagnetic minerals, and even sets within the latter, are possible through the use of specially designed electromagnets such as the Franz Isodynamic Separator.

Paramagnetic minerals include siderite, diopside, beryl, chalcopyrite, pyrite, and others, especially those which are iron-bearing. Diamagnetic minerals are calcite, zircon, wulfenite, and others. Research has shown that the oriented magnetic properties of minerals are analogous to the optical properties of the same paramagnetic mineral.

Taste: The sensation of taste can be determined only for those minerals which are easily soluble. Halite yields a salty taste, trona the taste of soda, epsomite a bitter taste, chalcanthite the taste of sulphuric acid and is nauseating. A cooling taste comes from niter and an astringent taste for alum. Glauberite has a bitter, salty taste, borates are salty, and borax-kernite has a sweet alkaline taste.

Odor: Except in rare cases minerals do not give off odors, yet if moistened with the breath, or in other cases rubbed or struck, some volatiles may escape and be detected by their odor. Some calcites, for example, when rubbed, and also some quartz, yield a very fetid odor, like rotten eggs. This may represent release of hydrogen sulfide which was originally trapped in small vacuoles in the minerals when they formed.

Kaolin, and other clay minerals, when moistened with the breath give off the odor of moistened clay. If arsenopyrite is rubbed vigorously by a harder object, a garlic odor is produced. Likewise pyrite and other sulfides give off a sulphurous odor when rubbed vigorously by a hard object. Heat will produce sulphurous fumes from sulfides, horse-radish odor from selenium-bearing ores, and garlic odor from arsenical compounds.

Electrical Properties: Some minerals, especially those of metallic luster, conduct electricity, while those of non-metallic luster generally are non-conductors. The isometric system shows equal conductivity in all directions and varies with the crystal system for other minerals.

Some minerals develop positive and negative charges on different parts of the same crystal when the temperature of the mineral is changed. These are called pyroelectric minerals. The phenomenon occurs only among the non-conductors. The best development is among the low-order symmetry minerals. Tourmaline is a good example with opposite charges developing at either end of the elongate crystal. Quartz shows a positive charge on heating on three alternate prisms and negative charges on the three other prismatic faces.

A piezoelectric quality has been noted for some minerals subject to pressure or tension. Quartz, for example, has excellent piezoelectric properties when subject to pressure or an electric field. This special characteristic lead to its use in radio transmitters during World War II. Well-developed, clear crystals were much sought after during this period. The development of synthetic quartz crystal grown in the laboratory caused a great drop in demand for the natural crystal.

Radioactivity: The radioactive decay of uranium and thorium toward their final lead isotope product is a process of emission of alpha particles, electrons and gamma ray radiation. The quantity of the daughter product lead to that of U or to Th is used to determine the age of the mineral. The radioactive emissions can be detected by special photographic film, but more commonly by instruments such as a geiger counter, a scintillator, or gamma ray spectrometer. The latter instrument identifies and quantifies daughter products of thorium and uranium and gives equivalent uranium and thorium content for the rock or ore samples. The radioactive isotope of potassium can also be measured. The other two instruments measure total radioactivity. With one of these relatively inexpensive and easily operated instruments you can determine the radioactivity and in turn identify those minerals which contain uranium, thorium, and in some cases, radioactive potassium in their composition. Instruments sensitive to radioactivity are widely used in the exploration for these minerals.

A group of about 150 minerals, some very rare, contain uranium or thorium as a part of their chemical composition. Some other minerals have radioactive elements in small quantities in solid solution. Based on economic value, uraninite and pitchblende, carnotite, tyuyamunite, coffinite, autunite, torbernite, brannerite, and davidite are the most important of the uranium minerals. Other minerals which contain smaller uranium content are quite numerous. Our select minerals of this latter group include monazite, the niobate-tantalates, allanite, zircon and some apatite. Thorium minerals of value are thorite and thorogummite and the oxide thorianite.

Fusibility: The relative degree of fusibility of minerals varies greatly among them. Some may fuse in the flame of a candle, many others are infusible even in the hottest flame. The test is made by holding a fragment in forceps and subjecting it to a hot flame. Please note that arsenical and antimonious compounds could injure the platinum plating in a high quality forceps set. The check for these two elements might best be made by an open or closed tube method, or charcoal block tests described below.

Fusibility is the tendency of a mineral to melt or become rounded on the edges of splinters when subject to the heat from a flame. For the test, fragment is held in the flame by forceps or a tweezer. Grades of fusibility range from 1-7. Fusion at 1-2 occurs in a candle flame, or the edges of a splinter are fused in the flame. If the mineral is readily fused in a bunsen burner flame, it is ranked (2). Fusibility of 3-6 is also observed with the use of a bunsen burner. A mineral which is readily fusible BB (before the blowpipe) has a fusibility of 3. When only the splinters and thin edges are readily rounded, BB fusibility is 4. When the splinters and edges are difficultly rounded, fusibility is 5. If only the tips of splinters and the thinnest edges show fusion BB with difficulty, fusibility is 6. When the mineral is not affected or is infusible, BB fusibility is rated at 7. Table 6 lists the fusibility rankings of various minerals according to the criteria above. Figure 22 shows the method of using the blowpipe and also the forceps method for holding the mineral fragment.

Tenacity: Tenacity includes brittleness, sectility, malleability and flexibility. Briefly defined brittleness refers to the mineral separating into a powder or grain when cut. Sectility is demonstrated when the mineral may be cut and not produce a powder or grains, though it may be powdered by hammering. Malleability refers to slices which may be cut off and flattened by hammering. Flexibility suggests the bending of a mineral wtihout breaking and remaining bent after the stress is removed. Elasticity refers to the bent mineral returning to its original shape and size. Mica illustrates the latter, and talc the former.

Streak: If a mineral is powdered from scratching its surface with a harder substance or by rubbing the mineral on an unglazed porcelain surface, the resulting powder color is called the streak. For many metallic luster minerals, this is an important diagnostic feature. Table 5 includes the characteristic streak of many minerals in our selected groups. It also shows the hardness of the mineral.

Colors of streak range from black to white and hardness refers to Mohs Scale. The colors of streak and hardness of a large number of minrals are included in the streak table. For best use of Table 5, first identify the streak color, then check for hardness. Using color of the streak where it intersects the hardness column should give the mineral or minerals having these characteristics. Additional reference to the mineral descriptions may permit a specific identification by these two simple tests.

TABLE 5. COLOR OF STREAK AND HARDNESS OF MINERALS

HARDNESS

STREAK	1	2	3	4	5	6
BLACK	Graphite (1.0-2.0) Wad (1.0) Pyrolusite (1.0-2.0)	Pyrolusite (2.0-2.5) Stephanite (2.0-2.5) Argentite (2.0-2.5) Shiny Jamesonite (2.0-3.0) Polybasite (2.0-3.0) Bournonite (2.5-3.0) Chalcocite (2.5-3.0)	Tetrahedrite (3.0-4.5) Stannite (3.5) Chalcopyrite (3.5-4.0) Enargite (3.0)	Niccolite (5.0-5.5) Skutterudite (5.5-6.0)	Wolframite (5.0-5.5) Ilmenite (5.0-6.0)	Magnetite (6.0) Columbite (6.0-6.5)
BLACK TO GREENISH BLACK	Molybdenite (1.0-1.5)		Millerite (3.0-3.5) Chalcopyrite (3.5-4.0)		Pyrite (6.0-6.5)	
DARK BROWN TO BLACK			Tetrahedrite (3.0-4.5)	Manganite (4.0)	Chromite (5.5) Wolframite (5.0-5.5) Uraninite (5.5) Ilmenite (5.0-6.0) Niccolite (5.0-5.5) Psilomelane (5.0-6.0) Hypersthene (5.0-6.0)	Columbite-Tantalite (6.3) Marcasite (6.0-6.5)
GREY BLACK	Covellite (1.5-2.0)	Stibnite (2.0) Galena (2.5) Chalcocite (2.5-3.0) Bismuthinite (2.0) Jamesonite (2.0-3.0) Bournonite (2.5-3.0)	Arsenic (3.5)	Iron (4.0-5.0)	Marcasite (6.0-6.5) Cobaltite (5.5) Smaltite (5.5-6.0) Arsenopyrite (5.5-6.0)	
LIGHT BROWN TO DARK BROWN	Bauxite (1.0-3.0)		Pentlandite (3.5-4.0)	Brannerite (4.5)	Chromite (5.5) Franklinite (5.5-6.5)	Cassiterite (6.0-7.0) Rutile (6.0-6.5)
REDDISH BROWN TO INDIAN RED		Pyrargyrite (2.5)	Cuprite (3.5-4.0) Descloizite (3.5)	Manganite (4.0)	Hematite (5.5-6.5) Franklinite (5.5-6.5) Ilmenite (5.0-6.0)	

TABLE 5. (CONT.) COLOR OF STREAK AND HARDNESS OF MINERALS

Streak Color							
BRIGHT RED	Hematite (1.0)	Cinnabar (2.0-2.5) Proustite (2.0-2.5)				Hematite (5.5-6.0)	Columbite-Tantalite (6.0)
DARK RED COPPER RED PURPLISH RED	Bauxite (1.0-3.0)	Cinnabar (2.0-2.5) Copper (2.5-3.0) Proustite (2.0-2.5) Pyrargyrite (2.5)					
PINK	Erythrite (1.5-2.5)						
YELLOW BROWN	Limonite (Earthy)		Sphalerite (3.5-4.0)		Limonite (5.0-5.5)	Goethite (5.0-5.5)	
LIGHT GREEN TO OLIVE GREEN GREENISH BROWN		Chlorite (2.0-2.5) Torbernite (2.0-2.5)	Atacamite (3.0-3.5) Brochantite (3.5-4.0) Malachite (3.5-4.0)	Brannerite (4.5)		Hornblende (5.0-6.0) Turquoise (5.0-6.0) Uraninite (5.5)	
GREY - LT. SILVERY, WHITISH BLUISH GREY GREY	Cerargyrite (1.0-3.0) Vivianite (1.5-2.0) Sylvanite (1.5-2.0)	Calaverite (2.5) Silver (2.5-3.0) Bismuth (2.5)	Arsenic (3.5) Antimony (3.0-3.5)	Platinum (4.0-4.5)	Uraninite (5.5)	Monazite (5.0-5.5) Brookite (5.5-6.0) Augite (5.0-6.0)	Epidote (6.0-7.0) Glaucophane (6.0-6.5) Cassiterite (6.0-7.0)
ORANGE TO YELLOW GOLD TO LIGHT YELLOW	Realgar (1.5) Sulphur (1.5-2.5) Carnotite (1.0-1.5) Bauxite (1.0-3.0) Orpiment (1.5-2.0)	Becquerelite (2.0-3.0) Crocoite (2.5-3.0) Gold (2.5-3.0) Autunite (2.0-2.5) Gummite (2.5-5.0)	Greenockite (3.0-3.5) Descloizite (3.5) Vanadinite (3.0) Gummite (2.5-5.0)	Zincite (4.0-4.5) Gummite (2.5-5.0)		Brookite (5.5-6.0)	Olivine (6.5-7.0)
LIGHT BLUE BLUE	Vivianite (1.5-2.0)		Azurite (3.5-4.0)	Sphalerite (3.5-4.0)		Lazurite	
WHITE	Bauxite (1.0-3.0) Cerargyrite (1.0-3.0) Sulphur (1.5-2.0) Vivianite (1.5-2.0)	Brucite (2.5) Ulexite (2.5) Chrysocolla (2.0-4.0) Glauberite (2.5-3.0) Gypsum (2.0) Gibbsite (2.5-3.5) Wulfenite (2.7-3.0) Serpentine (2.5-5.0) All sulfates except Jarosite (2.5-4.0)	Vanadinite (3.0) Mimetite (3.5) Pyromorphite (3.5-4.0) Cerussite (3.5-4.0) Rhodocrosite (3.5-4.5) Sphalerite (3.5-4.0) All Carbonates exc. Cu Siderite (3.5-4.0) Wavellite (3.5-4.0)	Smithsonite (4.0-5.0) Fluorite (4.0) Scheelite (4.5-5.0)		Anatase (5.5-6.0) Augite (5.0-6.0) Datolite (5.0-5.5) Cancrinite (5.5-6.0) Hornblende (5.0-6.0) Lazulite (5.0-5.5) Sphene (5.0-5.5) Turquoise (5.0-6.0) Rhodonite (5.5-6.5) Apatite (5.0)	Amblygonite (5.5-6.0) Spodumene (6.5-7.0) Diamond (10) Chrysoberyl (8.5)

TABLE 6. FUSIBILITY FOR SELECTED MINERALS

1	2	3	4	5	6	7
Antimony	Acmite	Actinolite	Andradite	Apatite	Anthophyllite	Alunite
Bismuth	Aegirite	Albite	Brochantite	Chlorite	Beryl	Anatase
Borax	Ambrygonite	Almandite	Diopside	Columbite	Biotite	Andalusite
Boulangerite	Anglesite	Analcite	Hedenbergite	Cordierite	Enstatite	Aragonite
Bournonite	Apophyllite	Anhydrite	Hypersthene	Jarosite	Hemimorphite	Bauxite
Calaverite	Argentite	Atacamite	Labradorite	Microcline	Margarite	Brookite
Carnallite	Arsenopyrite	Autunite	Nephelite	Orthoclase	Muscovite	Brucite
Cerargyrite	Axinite	Barite	Pyrope	Plagioclase	Phlogopite	Calcite
Cinnabar	Azurite	Bravoite	Sodalite	Scheelite	Serpentine	Carnotite
Epsomite	Bornite	Celestite	Spodumene	Siderite	Strontianite	Cassiterite
Kernite	Cancrinite	Chabazite	Tremolite	Sphalerite	Tourmaline	Chromite
Mimetite	Cerussite	Chalcanthite	Willemite	Tantalite	Uvarovite	Chrysoberyl
Orpiment	Chalcocite	Cobaltite	Wollastonite	Thorite		Chrysocolla
Pyrargyrite	Chalcopyrite	Copper				Corundum
Realgar	Colemanite	Covellite				Cristobalite
Soda Niter	Crocoite	Cuprite				Diamond
Stibnite	Cryolite	Enargite				Dolomite
Sulphur	Datolite	Epidote				Dumortierite
Sylvanite	Galena	Erythrite				Franklinite
Tellurium	Glauberite	Fluorite				Goethite
Ulexite	Halite	Glaucophane				Greenockite
	Heulandite	Gold				Hausmannite
	Lepidolite	Gypsum				Hematite
	Loellingite	Hornblende				Hydrozincite
	Malachite	Idocrase				Ilmenite
	Millerite	Jadeite				Kaolinite
	Natrolite	Lazurite				Kyanite
	Niccolite	Marcasite				Leucite
	Pectolite	Pyrite				Limonite
	Pentlandite	Pyrrhotite				Magnesite
	Polybasite	Rhodonite				Magnetite
	Prehnite	Silver				Manganite
	Pyromorphite	Skutterudite				Microlite
	Scapolite	Spessartite				Molybdenite
	Stephanite	Sphene				Monazite
	Stilbite	Torbernite				Olivine
	Sylvite	Wolframite				Opal
	Tennantite	Zoisite				Platinum
	Tetrahedrite					Psilomelane
	Vanadinite					Pyrolusite
	Vivianite					Pyrophyllite
	Witherite					Quartz
	Wulfenite					Rhodochrosite
						Rutile
						Sillimanite
						Smithsonite
						Spinel
						Staurolite
						Talc
						Topaz
						Turquoise
						Uraninite
						Zincite
						Zircon

Fluorescence: Some minerals fluoresce when exposed to the short-wave length of ultra-violet rays in the range of 2500 angstroms. (An angstrom unit is approximately 1/254,000,000 of an inch.) The ultraviolet rays from the sun are in the range of 2900 to 4000 angstroms and are invisible. The visible light rays are between 4900 and 8000 angstroms. Infrared rays which provide heat for the earth are in the range of 8000 to 25,000 angstroms.

All minerals are made up of atoms, and each atom is composed of electrons surrounding a nucleus of protons, neutrons and other particles. The protons of the nucleus have electrons revolving in definite orbits about the nucleus. The number of protons in the nucleus determines the number of electrons in orbit about it.

When ultra-violet rays strike the atom in the form of photons, they cause some electrons of the atoms to change orbital position around the nucleus. At the same time a small amount of energy in the form of heat is produced. When the electrons return to their original orbital position, the absorbed photons are re-radiated at a longer wave length. The wave length of the re-radiated light determines the color of the fluorescence of the substance.

The process of absorbing energy, electrons moving into new orbits, giving up energy and returning to their original orbital pattern takes place in a very small fraction of a second. We see only the resultant light when the mineral is exposed to the stimuli of the ultra-violet rays. This phenomenon is called fluorescence.

Some minerals are much slower in their reaction time, and the electrons remain in an unnatural orbit for an appreciable length of time. In such cases we continue to see the light after the ultra-violet light source has been removed. This phenomenon is called *phosphorescence*.

Some minerals which do not normally fluoresce will sometimes do so because of an activating agent. For instance, most forms of calcite do not fluoresce. Yet, if a small amount of manganese is present in the mineral, it acts as an activator and causes the calcite to fluoresce red. The brilliance of the fluorescence varies with the amount of manganese present. Uranium salts in various rocks or minerals are also activators and cause a green fluorescence. A mineral may be listed as fluorescent, but the fluorescence may result from a coating of a fluorescent substance. Thus, minerals may fluoresce from one locality and not from another due to the lack or presence of an activator or fluorescent coating. Table 7 lists minerals fluorescent if an activator is present. Table 8 lists the normal fluorescent behavior of various minerals to ultra-violet short wave length light.

TABLE 7. MINERALS NORMALLY NOT FLUORESCENT, BUT FLUORESCENT IF AN ACTIVATOR IS PRESENT

<u>Mineral</u>	<u>Color of Fluorescence</u>
Albite	Very slight
Aragonite	White, yellow, green, red, pink
Hemimorphite	Cream
Quartz	Brownish yellow
Topaz	Light green
Chalcedony	Green, yellowish green, blue green
Beryl	Green
Calcite	Pink to red to blue
Amethyst quartz	Deep blue
Apatite	White, yellow, blue, green
Zircon	Bright orange (variable)
Alunite	White

TABLE 8.   NORMAL FLUORESCENT REACTIONS OF VARIOUS MINERALS
TO ULTRA-VIOLET (SHORT WAVE LENGTH) LIGHT STIMULI

Color of Fluorescence	Color of Phosphorescence	Mineral
White	Bluish White	Colemanite
White		Amblygonite
Yellowish White		Barite
Gray		Dumortierite
White, Yellowish Green Pale Yellow Green		Dolomite
Bluish Gray		Glauberite
	Blue White	Celestite
White, Bright Light Blue, Yellowish		Scheelite
Pale Yellow		Diaspore
Yellow		Anglesite
Yellow		Apophyllite
Yellow		Tyuyamunite
Yellow		Wollastonite
Yellow		Witherite
Bright Yellow		Wernerite
Yellowish Green		Uranium Salts
Yellowish Green		Autunite
Light Green		Lepidolite
Green		Opal
Bright Green		Willemite
Deep Green		Gypsum
Green, Blue, Yellow		Cerussite
Bluish Green	Bluish Green	Strontianite
Blue Green, Blue, Whitish Blue, Orange		Fluorite
Rare	Greenish Blue	Borax
Light Blue		Diamond
Light Blue		Magnesite
Light Blue	Light Blue	Trona
Intense Light	Blue	Scheelite
Blue	Blue	Wavellite
Blue		Smithsonite
Blue		Feldspar
Pink		Tremolite
Red-Faint Green		Halite
Bright Red		Spinel
		Uranium Salts
Deep Red	Pale Yellow, Red, Brown,	Spodumene
Orange Red		Sodalite
Orange Red		Kunzite
Bright Orange (Variable)		Zircon
Bright Orange	Orange, Yellow, Light Green	Pectolite
Bright Orange		Sphalerite
Slight Dark Brown		Crocoite

CONFIRMATORY

CHEMICAL

TESTS

INTRODUCTION

This chapter of the book describes some basic chemical tests which are very useful in determining the specific identification of a mineral. The chemical tests are categorized under solubility, heating tests, and microchemical tests.

1. Solubility tests – The important and useful Table 9 lists the reaction of various minerals to water and acids. Certain minerals which effervesce when acid is applied are also shown.

2. Heating tests – A series of mineral tests under heating tests include:
    a) Blowpipe and charcoal block tests
    b) Use of oxidizing flame (OF) and reducing flame (RF) in mineral tests
    c) Fusibility tests
    d) Flame tests for minerals subject to an open flame
    e) Open (OT) and closed tube (CT) tests
    f) Bead tests

3. Microchemical analyses of minerals

Simple Preliminary Chemical Tests

The specific identification of a mineral may be much assisted through the use of some simple preliminary chemical tests where the physical properties of a mineral are not completely diagnostic.

Solubility Tests

Table 9 lists the reaction of certain minerals to water and some selected acids. The normal procedure is to first check for the solubility of a mineral fragment in water. If the mineral gradually disappears or dissolves in the water, it is considered soluble. It should fall within that category of the table.

If solubility is not visible in water, then a clear fragment of the mineral may be tested for solubility in 1:7 (one volume of acid to 7 parts water) HCl (hydrochloric acid). If bubbles arise in the liquid drop, the mineral is said to effervesce. The presence of effervescence indicates the generation of a gas, e.g., calcite ($CaCO_3$) effervesces vigorously in HCl acid and CO_2 (carbon dioxide) is generated and the remaining calcium element remains in solution. If the test is made on a rock and the mineral fragment is small, the action of a reactive effervescent mineral can be heard by bringing the specimen close to, but not touching the ear. The slight "pop-pop" sound confirms the effervescence.

Other stronger acids may be used on mineral fragments not reactive to water or 1:7 HCl. Minerals reactive to these acids and the character of the reaction are also shown in Table 9.

Care should be taken in using any of the reagents listed. They are vigorously reactive to clothes and skin and lips and eyes. Keep the reagents away from the body and skin. Remember that if you mix acid and water, never pour the water into the acid. The heat generated by the mixing may cause the acid to explode violently from the container. Always pour the acid very carefully into water. It sinks slowly through it because of its greater density and the heat of reaction is absorbed by the water. Because of the small quantities of acid used, this should pose no serious problem, but may if larger volumes are used.

TABLE 9. REACTION OF MINERALS TO
WATER & ACIDS

Soluble In Water

Borax
Chalcanthite
Epsomite
Glauberite (P)
Gypsum (P)
Halite
Niter
Soda Niter
Sylvite
Trona
Ulexite (P,H)

Effervesces In HCL

Aragonite
Azurite
Calcite
Cancrinite
Dolomite (H)
Hydrozincite
Magnesite (H)
Malachite
Rhodochrosite (H)
Siderite
Smithsonite
Strontianite
Witherite

Soluble in HCL

Actinolite (P)
Allanite (Gel)
Amblygonite (Sl)
Analcite (Gel)
Anhydrite (H)
Ankerite
Antlerite
Apatite
Aragonite
Apophyllite (Dec)
Atacamite
Aurichalcite
Autunite
Azurite
Boracite
Brucite
Carnotite
Cassiterite (P)
Chabazite
Chlorite (P)
Chrysocolla
Colemanite (H)
Cordierite (P)

Copper
Cuprite (Conc)
Datolite (Gel)
Dolomite (H)
Epidote
Erythrite (>Red Sol)
Glauberite
Greenockite (>H₂S)
Gypsum (H)
Hematite (P)
Huelandite
Hypersthene (P)
Hydrozincite (P)
Idocrase (P)
Ilmenite (H)
Lazurite (>H₂S)
Lepidolite (P)
Leucite (Dec)
Malachite
Magnesite (H)
Magnetite (H)
Marcasite
Monazite
Natrolite (Gel)
Nepheline (Gel)
Olivine (H)
Pectolite
Psilomelane (>Cl)
Pyrolusite
Pyrrhotite (>H₂S)
Scheelite (>Y Powd)
Serpentine (Dec)
Siderite (H)
Smithsonite
Sphalerite (>H₂S)
Sphene (P,H)
Stibnite
Strontianite
Sodalite (Gel)
Trona
Turquoise
Tyuyamunite
Vanadinite
Wavellite
Willemite (>Gel)
Witherite
Wolframite (P)
Zincite

Soluble in H₂SO₄

Alunite
Amblygonite (Powd)
Apatite
Atacamite
Biotite (>SiO₂)

Brucite
Carnotite
Cryolite (>HF)
Cordierite (P)
Fluorite
Microlite
Phlogopite (Dec)
Pyrophyllite (P)
Serpentine (Dec)
Sphene (Dec)
Spinel (Diff in conc)
Staurolite (P, Dec)
Sylvanite (Conc>Red sol)
Topaz (P)
Tyuyamunite
Uraninite
Wolframite (P)
Zincite
Zircon (Powd in Conc)

Soluble in HNO₃

Actinolite (P)
Anglesite (Diff)
Antlerite
Apatite
Arsenopyrite (>S)
Atacamite
Autunite
Azurite
Bismuthinite (H)
Bornite (>S)
Brucite
Calcite
Carnotite
Cassiterite
Cerussite (Efferv)
Chalcocite
Chalcopyrite
Cobaltite (H)
Copper
Cordierite (P)
Dolomite (H)
Galena
Hemimorphite
Magnesite (H)
Marcasite
Millerite
Mimetite
Molybdenite (Dec)
Natrolite
Niccolite
Proustite
Pyromorphite
Scheelite (>Y Powd)

Silver
Skutterudite
Sodalite
Stannite (S & SnO₂)
Stephanite (H)
Stibnite
Sylvanite (>Au)
Sylvite
Tetrahedrite
Torbernite
Turquoise
Tyuyamunite
Uraninite
Vivianite
Willemite
Zincite
Zircon

Soluble in Aqua Regia

Calomel
Enargite
Gold
Niccolite
Platinum (P)
Wolframite

EXPLANATION

P = Partially Soluble
H = Heating Required
Gel = Gelatinizes
Diff = Difficultly Soluble
Dec = Decomposes
> = Yields
Conc = Concentrated Acid
Sl = Slowly
Powd = Powder
Sol = Solution
Y = Yellow

Heating Tests

The blowpipe, charcoal block, plaster of Paris tablet, closed and open tubes, flame tests, fusibility, and bead tests are useful techniques in mineral determinations. These have long been used by mineralogists. A brief comment and description on each of these additional confirmatory tests follows.

Blowpipe tests: Several chemical and mineralogical tests are possible with the blowpipe. This is a small instrument commonly used for confirmatory mineral identification. An example of the blowpipe is shown in figure 21.

The source of flame and heat may be a bunsen burner or if in the field bottled gas, an alcohol lamp, or a stearin candle. The procedure involves blowing into the blowpipe with its tip in the edge of the flame (see figure 22). By breathing through the nose and extending your cheeks, a continuous supply of air can be supplied to the blowpipe and a very hot flame produced. The characteristics of the oxidation flame (OF) and reduction flame (RF) are shown in figure 22. The application of the RF or OF is indicated in the tests listed for the individual minerals.

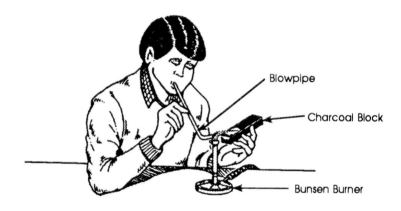

Figure 21. Use of blowpipe and charcoal block.

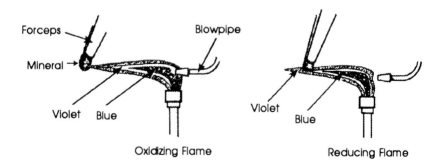

Figure 22. Schematic view of the oxidizing flame (OF) and the reducing flame (RF) using a blowpipe.

Charcoal block tests: The charcoal block is rectangular and about four inches long. A mineral fragment or powdered mineral is placed in a small cavity cut in the block near one end. It is then fused before the blowpipe flame. Before the blowpipe is abbreviated as BB in the mineral descriptions. Some volatile minerals produce distinctive sublimates on the charcoal block, others metallic globules in the strongly reducing environment. The reactions of certain minerals are noted in the individual mineral descriptions. Once the test is completed, a simple scraping and cleaning of the block by a knife blade after each test permits reuse of the charcoal block (figure 21).

Flame tests: Flame tests can be performed in several ways. A common method is the use of a set of special forceps to hold the mineral which is then placed in a hot open flame such as that from a blowpipe (see figure 22). Observe the color of the flame produced. This color relates to the reactive element in the mineral. If the mineral fragment is moistened with HCl (hydrochloric acid) or HNO_3 (nitric acid) before performing the flame test, a better flame color often results. Examples of elements which produce good flame color tests are barium–which produces a distinctive green yellow–or lithium which yields a deep red against a dark background.

If an abundance of the mineral is available, the powdered mineral may be sprinkled into the flame and the color observed. For small mineral fragments, the fragment may be placed in a small loop in a platinum wire which has previously been imbedded in a glass rod. The test is performed as above. Table 10 is a list of several flame colors expected from certain elements present in minerals.

TABLE 10. Expected Flame Colors for Certain Elements in Minerals

Light Blue	Arsenic
	Lead
	Tellurium
Deep Blue	Copper
	Selenium
Blue Green	Antimony
	Phosphorus
Green	Boron
	Copper
Green Yellow	Barium
Yellow	Sodium
Yellow Red	Calcium
Red	Lithium
Red Purple	Strontium
Violet	Potassium

Open tube test: Open tube tests are conducted by using a glass tube about 3/8 inch in outside diameter and about 5-6 inches in length. To prepare an open tube, heat the glass tube and turn it slowly in the flame. When it is red, bend it to the shape as shown in the sketch (figure 23). A small quantity of mineral can be placed in the open tube at the bend and then heated over the flame. The short end of the tube should be horizontal and the longer portion inclined upward. Check for odors from the tube and watch for sublimates which form on the inside of the tube. The mineral which you are attempting to confirm will indicate in its description if it is amenable to this type test. If it is, the results expected will be described.

Closed tube test: A closed tube test can be made by taking a 7-inch length of 1/4 inch outside diameter tubing, heating the center of the tubing to redness over a bunsen burner flame and when it is softened, pull it apart and close the ends through the reheating of the two parts. For use in mineral identification, the sample is heated in the closed end of the tube (figure 23). Odors and sublimates in the tube are noted and compared to those which apply to the mineral being tested.

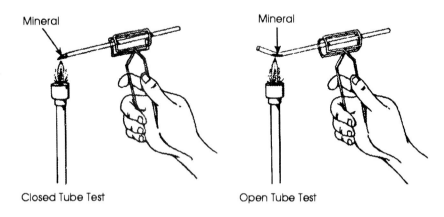

Closed Tube Test   Open Tube Test

Figure 23. Closed and open tube tests of minerals.

Bead tests: Several relatively simple fluxes are available for bead tests. To perform these, prepare a glass rod in which you imbed a platinum wire. This should be about one inch in length and have a small (about 1/8-inch diameter) loop at the end of it. Heat the wire loop and touch it in the desired flux. Borax, sodium carbonate, or salt of phosphorous are the common fluxes. Continue heating the loop and touching the flux until a good bead is formed in the loop. With the hot bead in the wire loop, touch the bead against a very small amount of the powdered mineral to be tested. Fuse this into the bead by reheating. Observe the resulting color change. The bead test may require more than one touch and fusion of the powdered material. Some distinctive reactions of certain elements to the simple fusion tests of borax and salt of phosphorous are given in Table 11 using the oxidizing and reducing flame.

Glass Rod   Platinum Wire

Loop for flux

TABLE 11. Bead tests with borax and salt of
phosphorus fluxes

Borax Bead Tests		
Color of Bead	Oxidizing Flame and Elements*	Reducing Flame and Element
Colorless, white, opaque	Si, Ca, Al, Zn, Fe-cold	Mn – also pinkish
Gray – milky		Ni
Red, reddish-brown, brown	Cr-hot, Mn-hot, Fe-hot, Ni-cold, Mn	Cu
Green	Cu-hot	Fe, Cr (Emerald green
Yellowish green	Cr-cold	U-concentrated
Yellow	Fe-cold and hot, U-cold and hot, Cr hot conc.	
Blue green	Cu-cold conc.	
Blue	Co-hot and cold, Cu-cold conc.	Co-hot and cold
Violet	Ni-hot, Mn-hot	
Salt of Phosphorus Bead Tests		
Colorless	Ti-cold	
Red	Cr-hot	V-Brownish red, hot

Green	Cr-cold
	Mo-yellowish green, hot
	U-yellow green, hot
	U-green, hot
		Cr-cold
		Mo-murky green, hot
		  clear green, cold
		U-green, cold
		V-green, chrome-
		  green, cold

Yellow	V-hot, cold, Mo-hot, cold Ti-hot	U-yellowish green Ti-hot
Violet	U-hot, V, Ti	Ti

* For elemental symbols see Table 2.

SYSTEMATIC MICROCHEMICAL ANALYSIS
(Opaque Minerals)

In rare cases the physical and other tests outlined above may not yield a clean cut mineral identification. The use of more sophisticated test procedures may permit a specific solution. For this reason we introduce the systematic microchemical technique as an additional confirmatory method of identification. While the method is mainly applicable to the opaque minerals, you will note several such tests can also be used for many of the non-metallic mineral varieties as well. The major elements of minerals which can be identified by the outlined procedures are shown in Table 12.

Microchemical Analyses of Minerals: A small, relatively inexpensive microchemical kit is suggested for your use. This kit will materially assist you in mineral identifications, and give many hours of enjoyment in your searches.

A list of materials and reagents which make up the kit is included in the following section. The kit can be compiled by you or supplied at a modest cost by MicroChem Industries.

TABLE 12. Mineral Elements Identifiable by Systematic Microchemical Analysis

Soft Minerals (Opaque)   (Minerals scratched by a needle)   By Residue: Ag, As, Bi, Cu, Fe, Ni, S, Sb, Se, Te, Pb, Au, Mg, Mn   By Solution: Ag, Co, Cu, Fe, Ni, Zn, Pb, Se, Te, Bi, Sb, As, Au, S, Sn
Hard Minerals (Opaque)   (Minerals not scratched by a needle)   By powder color and mineral possibilities: mainly iron-bearing oxides, cassiterite, tungstates, rare earth minerals.   By those difficultly soluble in concentrated HNO3, 1:1 HNO3 or Aqua Regia: Ni, W, Sn, Ta, Co, Nb.   By those soluble in HNO3 and/or Aqua Regia: Mainly Co-Fe-Ni arsenides, sulfarsenides, sulfantimonides, sulfides. Examples: pyrite, marcasite, niccolite and the elements: As, Bi, Co, Cu, Fe, Ni, Sb, S, Mn

Please note that the use of the simple chemical tests suggested in the earlier section of the book may give confirmation of the mineral identification before proceeding to this stage.

The basic equipment and supplies required for the microchemical tests are included in Table 13. Figure 25 illustrates a simple kit container for the necessary supplies and equipment. The kit is completely portable, and of relatively low cost. You can prepare it or arrange for others such as Micro-Chem Industries (see p. 234) to make it from the sketch shown. Chemical supplies can be secured from a chemical supply house.

You will note that in addition to the chemical reagents, some clean glass microscopic slides, a glass rod with an embedded platinum wire in one end, a small alcohol lap, a needle embedded in a wooden pencil-like rod and about 4 inches long, and some small capillary glass tubes with openings approximately 0.06 to 0.12 mm in diameter are required for the chemical tests. Most of the microchemical tests are carried out on glass slides. Hence, a good hand lens, but preferably a small binocular microscope is needed to observe the reactions.

The platinum wire (about .35 mm in diameter) is bent into a loop at its end to an inside diameter of about 0.62 mm. It is used to apply drops of the liquid reagent to the glass slide. It also doubles in performing bead and flame tests.

The scratch needle is embedded in a wood rod. The steel needle permits hardness determinations for Hard and Soft minerals. The alcohol lamp is used to heat the glass slides and the reagents on them, and to increase the rate of reaction of tested materials. The capillary tubes find use in transferring liquid drops from one part of the glass slide to another part. These also keep toxic reagents away from the mouth. The binocular microscope, or a good hand lens, is used to identify the crystal patterns and colors formed in the various chemical reactions. Some of these are specifically diagnostic of the element(s) which react with the reagent.

Systematic Microchemical Procedure

The microchemical procedures used for elemental and mineral determination are those outlined by M.N. Short of the U.S. Geological Survey (1940) for the opaque or metallic luster minerals. The great majority of the elements identified are based on the character of the crystals formed by precipitation within a liquid reagent on a glass slide. It follows that the larger the crystals formed, the easier the identification. Hence, conditions for crystal growth must be carefully controlled. In general, the reagent used to bring about the precipitation or color change in the drop should be more concentrated than the unknown elements in the drop to be tested. Too great a concentration of ions to form a precipitate may result in many small and poorly developed crystals. On the other hand, if too low a concentration is used, no precipitate will form. The sought after precipitate is somewhere between these two end points. Experimentation will yield the necessary concentration.

To give you experience and confidence in your ability to do the tests, we suggest that an experiment be performed by using a 0.5% solution of silver nitrate and a small fragment of ammonium bichromate. This solution and added small fragments of the ammonium bichromate will produce beautifully colored, well-developed, and readily recognizable crystals. Short also describes the use of 0.5% copper nitrate merged with a drop of potassium mercuric thiocyanate (3% solution in water). The reaction produces copper mercuric thiocyanate crystals of a very distinctive green color and shape. He concludes that these concentrations are about correct for most elements to be tested. If amorphous-like precipitates form, the concentration is probably too great. One can readily repeat the test with new material, or the drop can be diluted and transferred to a new place on the slide and then retested.

Short follows the two methods earlier described by chemists Chamot and Mason. Method I is most used and follows:

A drop of the reagent is placed near a drop of the solution to be tested. By means of a platinum wire or drawn-out glass rod (capillary tube), a tiny channel is made to flow from the reagent into the test drop. The slide can be slightly tilted to facilitate the flow. Under no conditions should the two drops merge completely.

The glass slide must be very clean to form the liquid channel. The sketch of figure 24 illustrates Method I of Chamot and Mason (1921).

TABLE 13. Reagents and supplies for the microchemical kit

Liquid Reagents (for plastic squeeze bottles)	Solid Reagents (capped vials)
Distilled Water 1:1 HNO_3 (1 part nitric acid in 1 part distilled water 1:7 HNO_3 (1 part nitric acid in 7 parts distilled water) HCl (concentrated hydrochloric acid) H_2SO_4 (concentrated sulfuric acid) 3% $K_2Hg(CNS)_4$ (3% solution of potassium mercuric thiocyanate Thiourea Dimethyl Glyoxime (2% in alcohol) NH_4OH; (concentrated ammonium hydroxide) Pyridine in HBr (Pyridine in hydrobromic acid) NH_4MoO_4 (ammonium molybdate) Sodium sulfide (saturated solution) Rhodomine Reagent	KI (potassium iodide) CsCl (cesium chloride) $SnCl_2$ (stannous chloride) $Co(NO_3)_2$ (cobaltic nitrate) KCNS (potassium thiocyanate) $K_4Fe(CNS)_6$ (potassium ferric thiocyanate) KNO_2 (potassium nitrite) $NaBiO_3$ (sodium bismuthate) RbCl (rubidium chloride) Zn (zinc) NH_4Cl (ammonium chloride) CaF_2 (calcium fluoride) NaCl (sodium chloride)

Supplies and Equipment

Alcohol Lamp (Metal or glass)	Matches
Glass Slides (25 x 75 mm)	Toothpicks
Capillary Tubes	Streak Plate
Dropper (plastic or rubber – glass)	Small Magnet
Forceps	Small Spatula
Glass Rod with platinum wire insert	Porcelain Test Plate (Opt.)
Test Tube Holder	Blowpipe
Test Tubes (3)	Charcoal Block 1" x 4"
Small Stearin Candle	Open Glass Tubes
Scratch Needle (mounted in wood rod)	Closed Glass Tubes

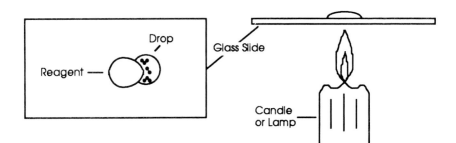

Figure 24. Merging of reagent and solution for a microchemical test of an element and slow heating of solution on glass slide.

Method II involves the use of a solid reagent and a drop of the solution of the substance to be tested. While frequently used, the method has the shortcoming of supplying too high a concentration of reagent ions, and it is not as easily controlled as liquids of known concentration. Short (1940) crushed the solid reagent to a small diameter size. With a distilled water moistened clean toothpick (not your tongue), he then transferred the smallest fragment visible by the unaided eye to the center of the drop. Particle size of the mineral tested is normally limited to fragments greater than 0.17 mm diameter. Fragments smaller in size may require additional means of identification such as x-ray, polarized light and/or reflected light. Additional reagent is added if necessary. The reaction is viewed under a binocular microscope to observe the results. For most of the outlined tests a good hand lens of at least 10 power can be used to identify the precipitates and their color.

Keep in mind that the tests suggested are not necessarily diagnostic of the mineral itself, but for some of the elements which make up the mineral. For example, a pink iron test may appear when chalcocite (Cu_2S) is dissolved in nitric acid and merged with potassium mercuric thiocyanate. This indicates the presence of some iron in the composition, or a microscopic iron-bearing mineral within the chalcocite such as pyrite. The presence of copper will be shown by distinct crystals in cross-like forms and/or prisms or moss-like aggregates which are normally of green color but are here tinted reddish. In another example, two different minerals may contain the same elements and will, of course, produce the same results. Hence in some cases the microchemical test may not singularly identify one mineral in preference to another. Other properties of the minerals used in conjunction with elemental composition, however, will permit identification.

Experimentation and experience with the use of the solutions and reagents will yield predictable results for your procedures and identifications. By taking care in reagent particle size, cleanliness of your glass slides, the rate and amount of heat applied in order to dissolve the mineral, the size of the reagent drop used, and other related factors, you can markedly influence the end results of the microchemical tests. Positive results are attainable. The important point is to take care in the steps of the method used.

The Systematic Microchemical Analysis outline for the opaque or metallic luster minerals (figure 26) is based on Short's earlier work (1940). It has been supplemented by Chace (1951) and later modified by Proctor (1971). In the figure two general analytical microchemical procedures are outlined. One of the methods applies to opaque "soft minerals" and the other to opaque "hard minerals." Soft and hard minerals are defined by Short as follows:

Soft Mineral: includes minerals which are scratched readily by a needle.
Hard Mineral: includes minerals which scratch with difficulty or not at all by a needle.

While more quantitative methods for mineral hardness of the opaque minerals are now available, these two group separations for mineral hardness, plus the experience of the individual, are most useful in this systematic procedure.

The procedure outlines the systematic microchemical analysis for the soft minerals by solution, residue and solution. Seventeen individual elements, mainly metallic in character, are quite readily identified by this procedure if one or more is present in the original substance. Always keep in mind that many of the reagents used are very toxic. Care must be taken to avoid any ingestion of the reagents into the mouth or onto the skin.

Procedure II outlines a systematic procedure for the determination of some hard minerals and their contained elements. Approximately 30 uncommon minerals and mineral groups and about 16 elements are potentially identifiable with this procedure. The possible identifiable elements and minerals and mineral groups of both procedures are summarized in Table 12.

We wish you well in this fascinating field of mineral study, description, and identification which will bring many hours of pleasant satisfaction to you.

48

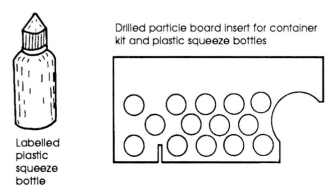

Labelled
plastic
squeeze
bottle

Drilled particle board insert for container
kit and plastic squeeze bottles

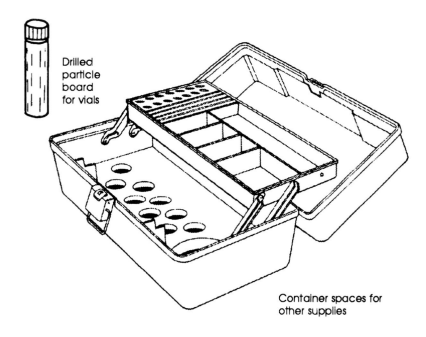

Drilled
particle
board
for vials

Container spaces for
other supplies

Figure 25. Microchemical kit for mineral determinations.

Figure 26 a. **Systematic Microchemical Analysis - I**
after Fred M. Chace with
Modifications by P.D. Proctor

SOFT MINERALS

GROUP A	Minerals soluble in 1:1 HNO_3.

PROCEDURE

Powder mineral. Place in 1:1 HNO_3. Warm gently and evaporate to dryness. If necessary, repeat with additional drops of HNO_3 and evaporations, each to dryness.

RESIDUE (For solution test of residue see next page)

Examine residue from original solution in plain light for indications of:
Ag, As, Bi, Cu, Fe, Ni, S, Sb, Se, Te.

1. White amorphous powder = S or Sb
2. Gelatinous ring = Ag or As
3. Curdy white structureless mass = Bi or Te
4. White lattice or boxwork texture = Pb (rarely Ag when in excess)
5. Light agglutinous mass may be S
6. Agglutinated brick-red mass = Se
7. Green residue = Cu or Ni
8. Yellow or orange residue = Fe

Note: Use confirmatory tests to prove presence of these elements.

CONFIRMATORY TESTS

Pb Confirmatory Test:
Leach original residue with two or three successive drops of H_2O.
Transfer and add to each drop a fragment of KI:

Lemon yellow hexagons and dislike plates = Pb

Au Confirmatory Test:
Add drop of aqua regia to original residue. Evaporate to dryness. Au is converted to orange-brown residue very soluble in H_2O. Leach residue with drop of H_2O. Transfer drop and merge in pyridine-double bromide solution:

Pleochroic red crystals = Au

Se and Te Confirmatory Tests:
CsCl and $SnCl_2$ tests: Leach original drop with 1:5 HCl. Transfer drop. Add reagent:

1. Merge drop in drop of 2% $SnCl_2$:
   Brick-red powder = Se

2. Add fragment CsCl:
   Lemon-yellow crystalline precipitate = Te

GROUP B	Soft minerals insoluble in HNO_3 but soluble in aqua regia. Probably contain Hg or Mn.

PROCEDURE

Test for Hg - Dissolve mineral in aqua regia. Evaporate to dryness. $HgCl_2$ residue forms lattice pattern. Mn forms a pink circular ring. Leach residue with 1% HNO_3. Transfer drop. Add fragment $Co(NO_3)_2$; when dissolved, add fragment of KCNS. Blue branching or acicular crystals indicate Hg. (Hg occurs only with Sb, Se, Te, S.)

Test for Mn - Dissolve mineral in aqua regia. Evaporate to dryness. Leach residue with 1:7 HNO_3. Transfer drop. Add fragment of sodium bismuthate. Solution turns pink or purple = Mn.

GROUP C	Soft minerals insoluble in both 1:1 HNO_3 and aqua regia.

No procedure given.

Figure 26 a. **Systematic Microchemical Analysis - I (cont.)**

SOFT MINERALS

SOLUTION TEST OF GROUP A RESIDUES

Step 1   Test for Ag, Co, Cu, Fe, Ni, Zn:

Leach residue from original mineral solution with 1% HNO_3. Transfer drop. Merge in drop of $K_2Hg(CNS)_4$:

1. Amorphous white powder = Ag
2. Indigo blue prisms = Co
3. Ppt. of green-yellow mass, prisms and crosses = Cu
4. Solution turns red = Fe
5. Small brown spherulites at margin of drop = Ni
6. White to gray feathery crosses = Zn

Step 2   Test for Ag and Pb:

Leach original residue with H_2O to remove HNO_3. Add drop 1:5 HCl:

1. Curdy white mass = Ag
2. Colorless needles = Pb

Transfer drop:

Step 3   Test for As, Cu, Se, Te:

Add fragment of KI to drop from Step 2:

1. Amorphous white, gray or yellow powder = Cu
2. Solution turns yellow near fragment = Bi or Sb
3. Fine amorphous chocolate brown ppt. = Se or Te

Note: All precipitates with KI are amorphous. Observe in reflected light: Se and Te precipitates will not be confused with As and Cu because of peculiar way Se and Te precipitate. When Se and Te are present distinguish two by CsCl and $SnCl_2$ confirmatory tests.

4. Cu and As ppts. are similar; As is yellower and sometimes orange. To confirm, use ammonium molybdate test for As on new mineral powder.

Step 4   Test for Au, Pb, Se, S:

Examine residue from original solution after being treated by Steps 1,2 and 3 in reflected light, using 16-mm objective:

1. Yellow metallic grains = Au   Check with
2. White PbCl needles = Pb   Confirmatory
3. Metallic red powder = Se   Tests[1]

Step 5   Test for As, Bi, Pb, Sb, Sn:

Fragment of CsCl is added to same drop in which KI test was made in Step 2. Restore drop to original size by adding 1:5 HCl:

1. Iodide ppt. of Step 2 will form orange amorphous ppt. at border of drop = As
2. Red hexagons = Bi
3. Orange hexagons = As or Sb   Check with
4. Colorless octahedrons = Sn   Confirmatory
5. Milky white crosses = Pb   Tests[1]

Figure 26 b.   **Systematic Microchemical Analysis - II**

HARD MINERALS

GROUP A   Minerals insoluble or very difficultly soluble in conc. HNO_3, 1:1 HNO_3 or aqua regia. Group consists of iron-bearing oxides, cassiterite, rare earth minerals.

PROCEDURE

Remove powder from polished section with a chisel-pointed needle.

Observe powder:

1. Nonmetallic red powder = hematite or lepidocrocite
2. Brick-red to light brown powder = goethite, chromite, cassiterite, huebner-ite, manganean columbite-tantalite
3. Dark-brown powder = franklinite, chromite, braunite, ilmenite, huebnerite, tantalite-columbite, maghemite
4. Black magnetic powder = magnetite, magnesioferrite, trevorite, iron, nickel-iron

Test for iron:

Take powder into solution by several applications of aqua regia followed by evaporation(s) to dryness. If orange to red powder is obtained on evapora-tion, test for iron by taking residue into solution with 1:7 HNO_3. Add fragment $K_4Fe(CN)_6$. Prussian blue precipitate = Fe.

Test for Ni, W, Sn, Ta, Co, Nb, etc., by using specific procedure for individual minerals.

GROUP B   Minerals soluble in 1:1 HNO_3 and/or aqua regia. Group consists of Co-Ni-Fe arsenides, antimonides, sulfarsenides, sulfantimonides,sulfides, and includes pyrite, marcasite and niccolite. (Niccolite is recognized by color)

PROCEDURE

Hard white minerals soluble in HNO_3 contain only As, Bi, Co, Cu, Fe, Ni, Sb, S. Dissolve mineral in conc. HNO_3 or aqua regia.
Observe residue:

1. White amorphous powder = Bi, Sb, S
2. Gelatinous ring around drop = As
3. Green residue = Ni
4. Yellow or orange residue = Fe
5. Pink residue = Co

Note: 1 to 5 are indications only; therefore they must be confirmed.

Figure 26 b. **Systematic Microchemical Analysis - II (cont.)**

HARD MINERALS

Test for As, Bi, Cu, Co, Fe, Ni, Sb, S.

PROCEDURE

Leach residue with 1% HNO_3:

SOLUTION: As, Co, Cu, Fe, Ni go into solution.

Transfer drop. Add 1% HNO_3. Evaporate almost to dryness. Add $(NH_4)MoO_4$ solution and slowly evaporate to dryness and then add a drop of 1% HNO_3:

Observe precipitate:
1. Minute yellow octahedrons or twinned isometric crystals = As

Filter solution from As crystals. Transfer drop. Add $K_2Hg(CNS)_4$:

Observe precipitate:
1. Solution turns red (faint color) = Fe
2. Blue acicular crystals = Co
3. Brown spherulites at edge of drop = Ni
4. Greenish yellow, mossy dendrites of cigar—shaped crystals in radiating groups = Cu

---

RESIDUE: Bi, Sb, S remain in original residue.

Leach original residue with 3 drops of 1% HNO_3 to remove trace of As. Then leach with H_2O to remove HNO_3. Add 1:5 HCl and transfer drop. Add fragment of KI:

Observe precipitate:
1. Amorphous yellow to orange ppt. = As
2. Solution turns light yellow = Bi or Sb (or both)

Add fragment of CsCl to solution:

Observe Precipitate:
1. Orange hexagons = As or Sb (must be confirmed - use cesium double iodide test)
    Note: Only one hard mineral contains Sb, ullmannite and var. willyamite and kallilite.
2. Red hexagons = Bi
3. No test is given for S (See Short's text for test)

GROUP C Minerals insoluble or slightly soluble in HNO_3, but readily soluble in aqua regia and 1:1 HCl. Group includes the hard manganese oxides. Test for Mn.

PROCEDURE:

Sodium bismuthate test for Mn: Take powder into solution by successive leaching with 1:1 HCl or aqua regia. Leach the residue with 1:7 HNO_3. Transfer the drop. Add fragment solid sodium bismuthate: Drop turns pink or purple=Mn.

TABLES FOR THE DETERMINATION OF NON-METALLIC AND METALLIC MINERALS

Table Arrangement

Color

Cleavage

MOHS Hardness

Specific Gravity

Light: (0-3.0)

Medium: (3.0-4.0)

Dense: (4.0-5.0)

Very Dense: (>5.0)

Luster   ●━━━━●   Non-Metallic

▬▬▬▬   Metallic

(Connecting line or block length indicates range of HARDNESS.)

INSTRUCTIONS

Determine the following mineral properties in order

1. Color

2. Cleavage or indistinct cleavage

3. Hardness

4. Specific gravity

5. Luster

6. Identify mineral on chart

7. For confirmation of specific mineral refer to detailed description and additional tests.

WHITE TO LIGHT GREY, COLORLESS, SILVERY

INDISTINCT CLEAVAGE

MOHS HARDNESS

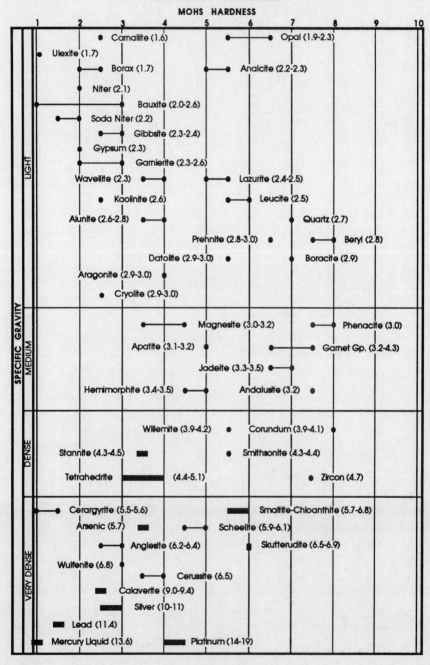

WHITE, LIGHT GRAY, COLORLESS, SILVERY
DISTINCT CLEAVAGE
MOHS HARDNESS

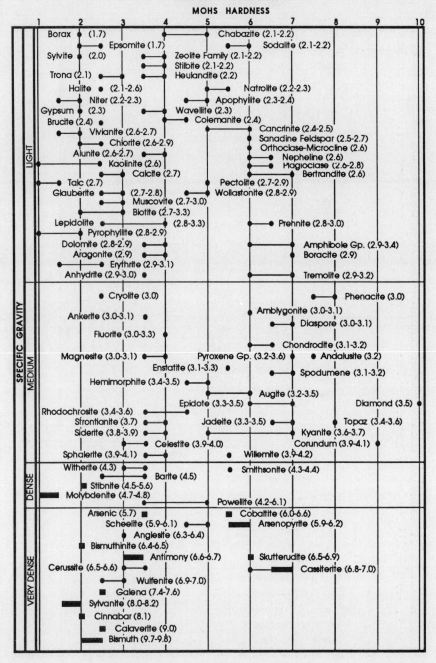

YELLOW, ORANGE, CREAM, GOLD, BRASS, BRONZE

INDISTINCT CLEAVAGE

MOHS HARDNESS

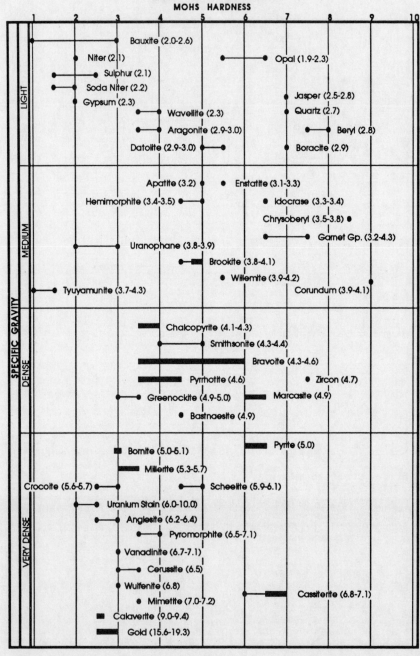

YELLOW, ORANGE, CREAM, GOLD, BRASS, BRONZE

DISTINCT CLEAVAGE

MOHS HARDNESS

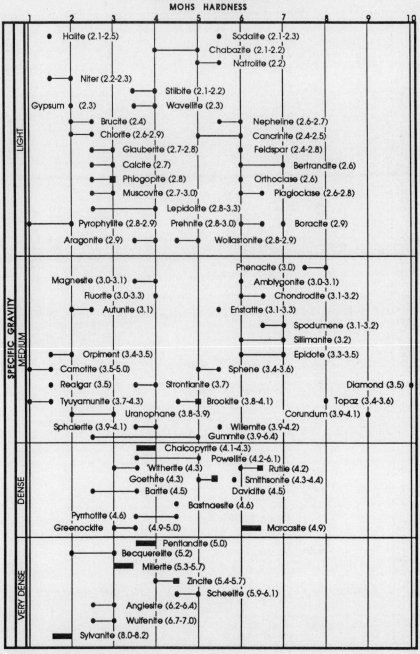

RED, PINK, COPPERY

INDISTINCT CLEAVAGE

MOHS HARDNESS

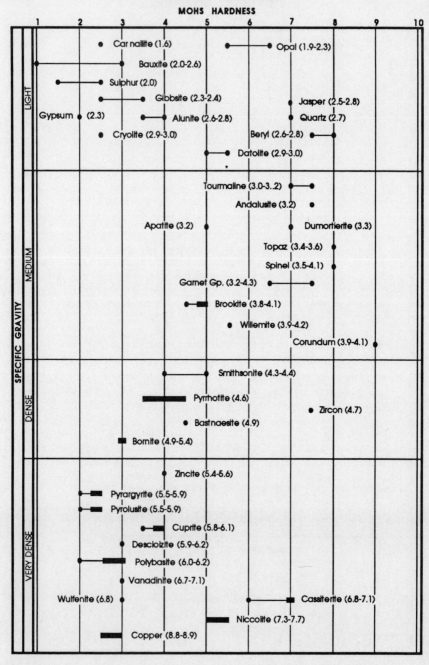

Carnallite (1.6)
Opal (1.9-2.3)
Bauxite (2.0-2.6)
Sulphur (2.0)
Gibbsite (2.3-2.4)
Jasper (2.5-2.8)
Gypsum (2.3)
Alunite (2.6-2.8)
Quartz (2.7)
Cryolite (2.9-3.0)
Beryl (2.6-2.8)
Datolite (2.9-3.0)

Tourmaline (3.0-3..2)
Andalusite (3.2)
Apatite (3.2)
Dumortierite (3.3)
Topaz (3.4-3.6)
Spinel (3.5-4.1)
Garnet Gp. (3.2-4.3)
Brookite (3.8-4.1)
Willemite (3.9-4.2)
Corundum (3.9-4.1)

Smithsonite (4.3-4.4)
Pyrrhotite (4.6)
Zircon (4.7)
Bastnaesite (4.9)
Bornite (4.9-5.4)

Zincite (5.4-5.6)
Pyrargyrite (5.5-5.9)
Pyrolusite (5.5-5.9)
Cuprite (5.8-6.1)
Descloizite (5.9-6.2)
Polybasite (6.0-6.2)
Vanadinite (6.7-7.1)
Wulfenite (6.8)
Cassiterite (6.8-7.1)
Niccolite (7.3-7.7)
Copper (8.8-8.9)

SPECIFIC GRAVITY

LIGHT

MEDIUM

DENSE

VERY DENSE

RED, PINK, COPPERY

DISTINCT CLEAVAGE

MOHS HARDNESS

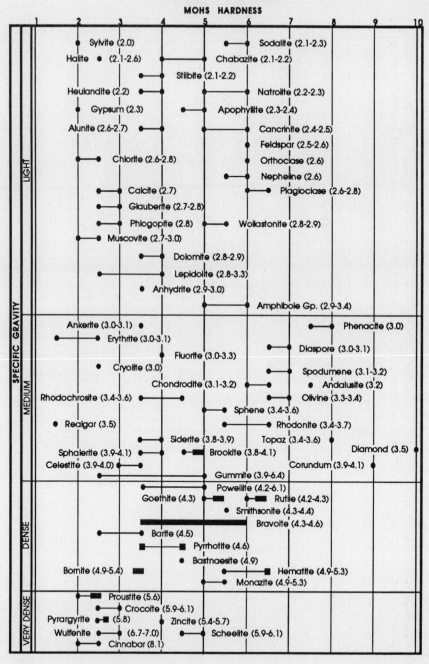

GREEN
INDISTINCT CLEAVAGE
MOHS HARDNESS

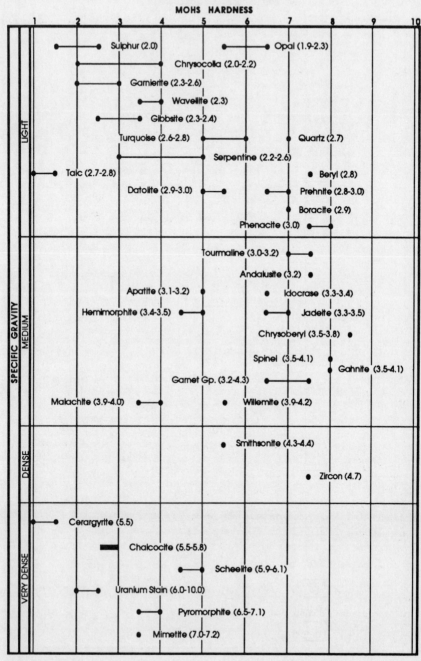

SPECIFIC GRAVITY

LIGHT

Sulphur (2.0)
Opal (1.9-2.3)
Chrysocolla (2.0-2.2)
Garnierite (2.3-2.6)
Wavellite (2.3)
Gibbsite (2.3-2.4)
Turquoise (2.6-2.8)
Quartz (2.7)
Serpentine (2.2-2.6)
Talc (2.7-2.8)
Beryl (2.8)
Datolite (2.9-3.0)
Prehnite (2.8-3.0)
Boracite (2.9)
Phenacite (3.0)

MEDIUM

Tourmaline (3.0-3.2)
Andalusite (3.2)
Apatite (3.1-3.2)
Idocrase (3.3-3.4)
Hemimorphite (3.4-3.5)
Jadeite (3.3-3.5)
Chrysoberyl (3.5-3.8)
Spinel (3.5-4.1)
Gahnite (3.5-4.1)
Garnet Gp. (3.2-4.3)
Malachite (3.9-4.0)
Willemite (3.9-4.2)

DENSE

Smithsonite (4.3-4.4)
Zircon (4.7)

VERY DENSE

Cerargyrite (5.5)
Chalcocite (5.5-5.8)
Scheelite (5.9-6.1)
Uranium Stain (6.0-10.0)
Pyromorphite (6.5-7.1)
Mimetite (7.0-7.2)

GREEN

DISTINCT CLEAVAGE

MOHS HARDNESS

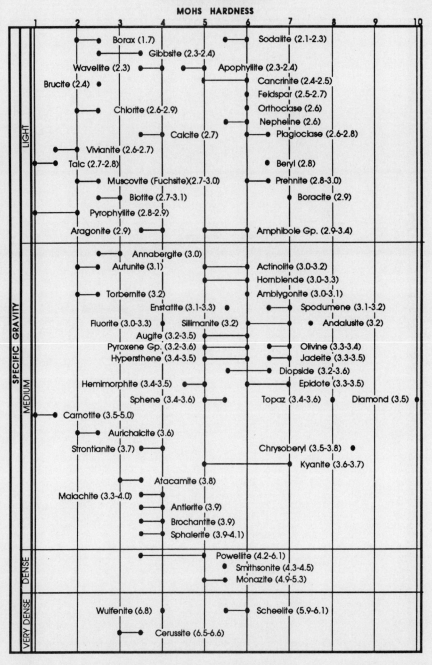

BLUE, PURPLE, VIOLET

INDISTINCT CLEAVAGE

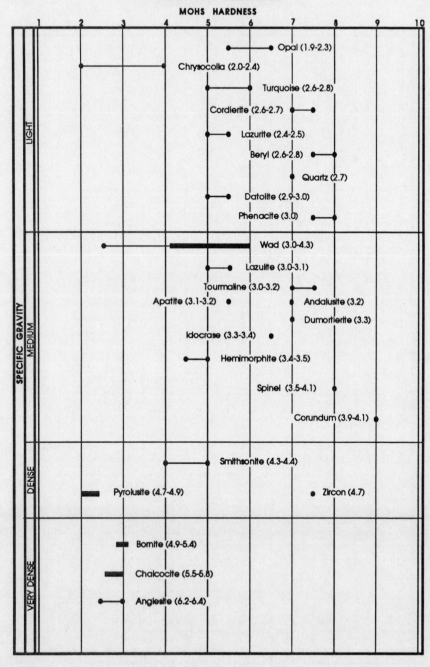

MOHS HARDNESS

Opal (1.9-2.3)
Chrysocolla (2.0-2.4)
Turquoise (2.6-2.8)
Cordierite (2.6-2.7)
Lazurite (2.4-2.5)
Beryl (2.6-2.8)
Quartz (2.7)
Datolite (2.9-3.0)
Phenacite (3.0)
Wad (3.0-4.3)
Lazulite (3.0-3.1)
Tourmaline (3.0-3.2)
Apatite (3.1-3.2)
Andalusite (3.2)
Dumortierite (3.3)
Idocrase (3.3-3.4)
Hemimorphite (3.4-3.5)
Spinel (3.5-4.1)
Corundum (3.9-4.1)
Smithsonite (4.3-4.4)
Pyrolusite (4.7-4.9)
Zircon (4.7)
Bornite (4.9-5.4)
Chalcocite (5.5-5.8)
Anglesite (6.2-6.4)

SPECIFIC GRAVITY — LIGHT · MEDIUM · DENSE · VERY DENSE

BLUE, PURPLE, VIOLET

DISTINCT CLEAVAGE

MOHS HARDNESS

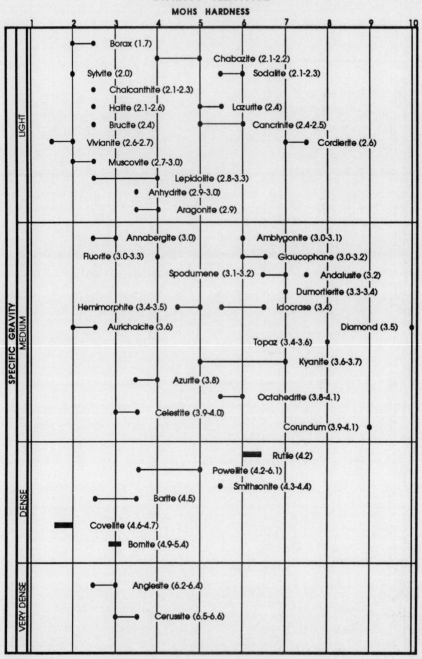

BROWN
INDISTINCT CLEAVAGE
MOHS HARDNESS

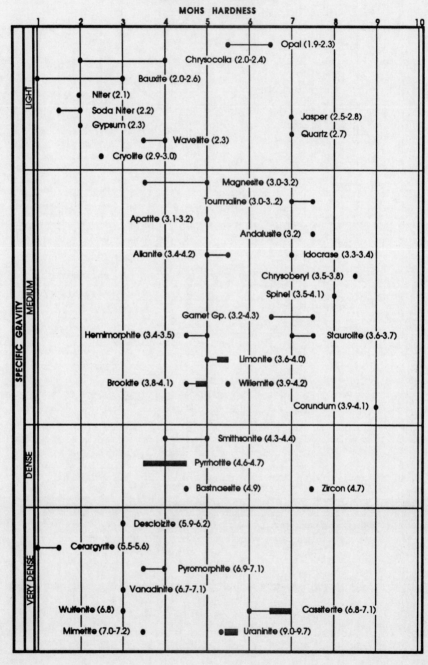

BROWN

DISTINCT CLEAVAGE

MOHS HARDNESS

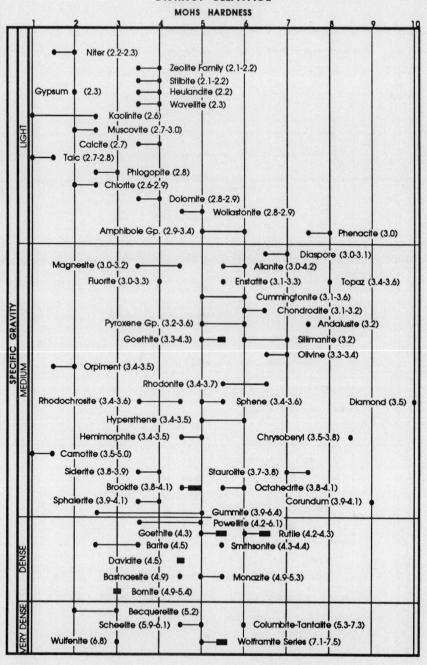

BLACK, DARK GREY

INDISTINCT CLEAVAGE

MOHS HARDNESS

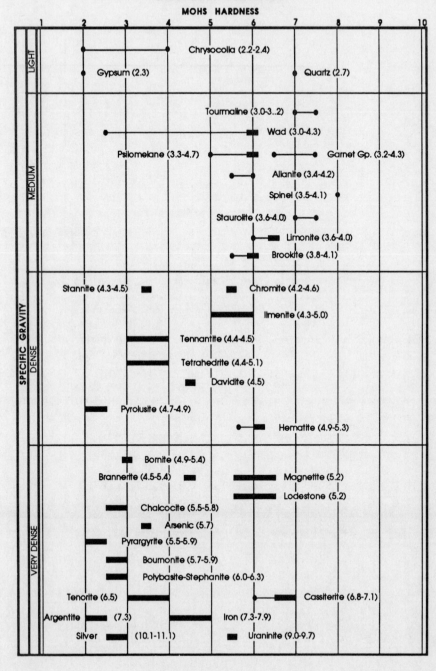

Chrysocolla (2.2-2.4)

Gypsum (2.3)   Quartz (2.7)

Tourmaline (3.0-3..2)

Wad (3.0-4.3)

Psilomelane (3.3-4.7)   Garnet Gp. (3.2-4.3)

Allanite (3.4-4.2)

Spinel (3.5-4.1)

Staurolite (3.6-4.0)

Limonite (3.6-4.0)

Brookite (3.8-4.1)

Stannite (4.3-4.5)   Chromite (4.2-4.6)

Ilmenite (4.3-5.0)

Tennantite (4.4-4.5)

Tetrahedrite (4.4-5.1)

Davidite (4.5)

Pyrolusite (4.7-4.9)

Hematite (4.9-5.3)

Bornite (4.9-5.4)

Brannerite (4.5-5.4)   Magnetite (5.2)

Lodestone (5.2)

Chalcocite (5.5-5.8)

Arsenic (5.7)

Pyrargyrite (5.5-5.9)

Bournonite (5.7-5.9)

Polybasite-Stephanite (6.0-6.3)

Tenorite (6.5)   Cassiterite (6.8-7.1)

Argentite (7.3)   Iron (7.3-7.9)

Silver (10.1-11.1)   Uraninite (9.0-9.7)

SPECIFIC GRAVITY

LIGHT · MEDIUM · DENSE · VERY DENSE

BLACK, DARK GREY

DISTINCT CLEAVAGE

MOHS HARDNESS

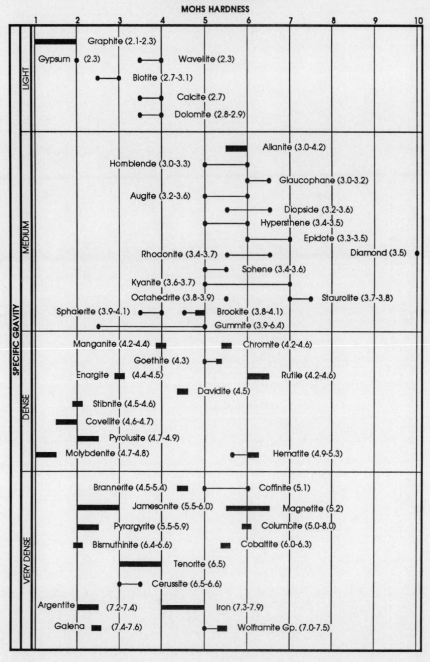

MINERAL
DESCRIPTIONS

with color photographs

The Victoria Diamond, in the rough, Natural Size

(Eng. & Min. Journal, 1887)

MINERAL DESCRIPTIONS

Explanation of Abbreviations and Terms Used

Bead – Fusion of special flux and mineral

BB – Abbreviation for before the blowpipe

Brittle – Easily fractured in fragment when cut by knife

CT – Closed tube test. See text

Elemental symbols – Abbreviation and special one or two letter symbol representing an element, e.g., Ag=silver. See Table 2A and 2B

Flame test – Burning of mineral in open flame for color test

Fluorescence – Reaction of mineral to short ultra-violet rays. See text discussion

Fuse – To melt a mineral before an open flame. See text and Table 6

Hardness – Relative hardness of mineral compared to other minerals. See Mohs hardness scale

HCl – symbol for hydrochloric acid

HNO_3 – Symbol for nitric acid

H_2SO_4 – Symbol for sulfuric acid

Luster – Light reflectance of mineral: metallic, non-metallic

MT – Microchemical test

MT 1-A – Michrochemical test on procedural charts, see page 49

OF – Oxidizing flame of blowpipe. See blowpipe section

Opaque – Non transmittal of light through mineral

OT – Open tube test. See text

RF – Reducing flame of blowpipe. See blowpipe section

Soluble – A mineral which can be dissolved in water and/or acids

Specific gravity –The weight of a mineral divided by the weight of an equal volume of water

Sectile – Cutability of a mineral with a knife without going to powder

Translucent – Light transmission, but objects not clear through mineral

Transparent – Light readily transmitted, object visible through mineral

ANTIMONY
Sb
Native Metal
R-Hexagonal

Perfect basal cleavage, fracture uneven, brittle. Massive, kidney-shaped lamellar Very light gray, opaque, metallic luster. Streak is tin-white. Hardness 3-3.5 and specific gravity is 6.6-6.7. Fuses easily BB, very volatile, yielding white sublimate. Occurs with silver, arsenic in veins associated with minerals stibnite, sphalerite, galena and pyrite. M.T. 1-A.

ARSENIC
As
Arsenic Ore
R-Hexagonal

One direction perfect cleavage, but rarely visible, uneven fracture. Rare crystals, massive, granular, botryoidal. Tin white tarnishes dark gray. Hardness 3.5, is brittle and has a specific gravity of 5.7. Metallic luster, opaque and has gray streak. BB volatilizes without fusing, on charcoal gives white coating and distinct garlic odor. In closed tube yields black band above mineral and silver gray band below. Rare mineral in veins associated with silver, nickel or cobalt minerals. M.T. 1-A.

BISMUTH
Bi
Ore Mineral of Bismuth
R-Hexagonal

Perfect basal cleavage, sectile, brittle. If warmed may be malleable. Silver white, opaque, metallic luster. Massive, granular, moss-like. Tarnished is reddish. Hardness is 2-2.5 and specific gravity 9.7-9.8. Streak silver white. BB fuses easily and volatilizes yielding sublimate of orange-yellow hot, and lemon-yellow cold. Soluble in HNO_3. Occurs in veins with quartz often with cobalt, nickel, silver and tin. M.T. 1-A.

DIAMOND
C
High Value Gemstone
Isometric

Octahedral perfect cleavage. Octahedral, dodecohedrons (Brazil), rarely cubes, faces often curved. Colorless, but brown, red, yellow, sometimes black and adamantine luster. Transparent. White streak. Extreme hardness of 10, specific gravity 3.5. Major host rock is kimberlite in pipes, also in river and beach gravels. Latter carry heavy resistate minerals as magnetite, garnet, monazite, sphene and others. Insoluble and infusible, may fluoresce blue and other colors. Important localities in South Africa, Brazil, India, Australia, Shades of Pangea, U.S.S.R. and even the U.S. in Arkansas and Wyoming. Remember the great diamonds: The Pitt, Hope, Oploff and Great Mogul all from India, and the Victoria and Tiffany from South Africa.

GRAPHITE
C
**Valuable Element When Pure
and in Quantity**
R-Hexagonal

Basal perfect cleavage. Flat tabular crystals, generally massive, foliated. Black with dull metallic luster, black streak. Sectile, very soft (1-2), greasy feel, flexible but not elastic. Specific gravity 2.1-2.3. Infusible and insoluble. Easily marks paper. Fairly common in veins, metamorphic, igneous and sedimentary rocks, associated with garnet, pyroxene, amphibole, chlorite, micas, and others.

ANTIMONY

ARSENIC

GRAPHITE

BISMUTH

DIAMOND

5 cm

COPPER
Cu

Copper Ore Mineral

Isometric

No cleavage and hackly fracture. Branching groups, scales, plates, wirelike, massive. Copper red, may have black tarnish. Has metallic luster, dull when tarnished. Opaque. Hardness is 2.5-3 and is malleable. Specific gravity 8.8-8.9. BB fuses at 3 to malleable globule. Later, when cool, has a black coating. Dissolve in HNO_3 and add sufficient NH_4OH and solution turns blue (Cu). Uncommon in large quantities. Occurs in oxidized zone of copper deposits, also in openings in igneous rocks, in veins and in pore spaces in sandstones and conglomerates. Associates are malachite, azurite, chalcocite, calcite, zeolites, native silver. M.T. 1-A.

GOLD
Au

A Very Precious Metal

Isometric

No cleavage, hackly fracture. Rare crystals, often distorted, nuggets, grains, wires, scales. Golden yellow to silvery yellow. Metallic luster, sectile and malleable. Very heavy, specific gravity 15.6-19.3. Hardness 2.5-3. Streak: gold yellow. BB fuses at 2.5-3, soluble in aqua regia. Associates are pyrite, chalcopyrite, arsenopyrite, marcasite, galena, sphalerite, quartz. M.T. 1-B.

IRON
Fe

Native Element
(rare)

Isometric

Cleavage, hackly fracture. Rare crystals, scattered grains to large masses. Steel gray to black, often rusty, metallic luster, steel gray streak. Malleable, with 4-5 hardness and 7.3-7.9 specific gravity. Strongly magnetic. Uncommon in earth rocks, in some meteorites. May be present in volcanic rock which cut coal seams. Has pitted or melted surface in meteorites. Nickel common associate in metallic meteorites, also in lesser quantity in stony and iron-stone meteorites. M.T. 1-A.

LEAD
Pb

Native Metal

Isometric

No cleavage, very malleable, ductile. Rare crystals, thin plates, rounded. Lead gray, opaque, metallic luster. Specific gravity 11.4 and hardness is 1.5. Dissolves easily in dilute HNO_3. BB fuses easily, yields yellow to white oxide on charcoal. Very rare mineral where it occurs in veins and placers in U.S., Mexico, Russia and Sweden. M.T. 1-A.

NATIVE COPPER

GOLD IN QUARTZ

WIRE GOLD

COPPER AND CALCITE

5 cm

NATIVE IRON

NATIVE LEAD

MERCURY	**Native Metal Fluid**	**Hexagonal**
Hg		**(Below -40°C)**

Occurs as small metallic globules. Tin white color, metallic shining luster, opaque. Fluid. Specific gravity is 13.6. Dissolves in HNO_3. Very volatile BB, very toxic in vapor form. A rare mineral associated with cinnabar, usually with former hot-springs and in volcanic areas. M.T. 1-A, 1-B.

SILVER	**Ore Mineral of Silver**	**Isometric**
Ag		

No visible cleavage, hackly fracture and malleable. Rare crystals, cubic or octahedral, branching, irregular plates, scales, wire-form. Silver white, tarnishes gray to black. Opaque and has metallic luster. Hardness is 2.5-3 and specific gravity 10-11. Fuses at 2 on charcoal and yields bright malleable globule. Latter dissolves in HNO_3; add HCl to solution and curdy white precipitate forms. Rare, occurs in veins, mainly in oxidized portion associated with other silver minerals, cobalt, uranium and arsenides. M.T. 1-A.

PLATINUM	**Ore Mineral of Platinum**	**Isometric**
Pt		

No cleavage visible, hackly fracture and malleable. Rare crystals, more commonly in grains and scales. Silver white to gray white, metallic luster and opaque. Shiny gray streak. Hardness is 4-4.5 and specific gravity 14-19. Sometimes magnetic, infusible. Insoluble in acid except hot aqua regia. Rare mineral element, found in placers and ultramafic igneous rocks.

SULPHUR	**Element**	**Orthorhombic**
S		

Poor cleavage, conchoidal to uneven fracture, brittle. Pyramidal, tabular, stalactic, massive. Yellow, with white streak. Transparent to translucent, resinous luster. Hardness 1.5-2.5 and specific gravity of 2.1. Fuses at 1, melts, burns readily with sulphur odor and blue flame. Volcanic setting as incrustations, in sedimentary rocks, mainly limestones. Associates: calcite, gypsum, anhydrite, aragonite, clay.

MERCURY with CINNABAR

SILVER

PLATINUM

SULPHUR

ORPIMENT   Arsenic Ore Mineral   Monoclinic
As_2S_3

  Perfect one direction cleavage, somewhat sectile. Crystals rare, foliated, columnar. Lemon yellow to brownish. Translucent and resinous to pearly on cleavage planes. Streak is yellow, hardness 1.5-2 and specific gravity 3.4-3.5. Fuses at 1, yields white coating BB on charcoal. In closed tube gives orange red sublimate. Associated with realgar, hot spring deposits, some gold deposits. M.T. 1-B.

REALGAR   Ore Mineral of Arsenic   Monoclinic
AsS

  Fair pinacoidal cleavage, conchoidal fracture. Rare crystals, massive, granular, crusts, earthy. Orange-red, resinous luster, transparent to translucent. Red to orange streak. Hardness is 1.5, slightly sectile; and specific gravity 3.5. Fuses at 1. BB on charcoal gives white coating and garlic odor. In closed tube and heated yields orange-red sublimate, yellowish toward open end. Uncommon, in some veins associated with orpiment and arsenic bearing minerals, often hot spring deposits. In light gradually changes to latter mineral. M.T. 1-A.

STIBNITE   Antimony Ore Mineral   Orthorhombic
Sb_2S_3

  Perfect one direction cleavage parallel to length of crystals, uneven fracture, slightly sectile. Prismatic striated crystals, curved, radiating, bladed, rarer massive, granular. Dark gray, lead gray, tarnishes black. Gray black streak. Metallic luster and opaque. Hardness 2 and specific gravity 4.5-4.6. Fuses at 1. BB mineral melts, yields sulphur odor and white coating on charcoal (Sb). Open tube produces sulphur odor and white sublimate in bottom of tube. Occurs in some quartz veins, replacement deposits in limestone, shale and hot spring deposits. Associates are galena, sphalerite, pyrite, cinnabar, realgar, orpiment, barite, sometimes gold. M.T. 1-B.

BISMUTHINITE   Bismuth Ore Mineral   Orthorhombic
Bi_2S_3

  One direction perfect cleavage. Rare acicular, mainly massive, fibrous, foliated. Lead gray, metallic luster and opaque. Hardness is 2, slightly sectile, streak is lead gray. Specific gravity: 6.4-6.6. BB fuses at 1 to round globule. Half and half potassium iodide and sulphur mixed with mineral and heated BB on charcoal gives a yellow coating next to mineral and bright red coating farther away indicative of bismuth. Soluble in HNO_3. This is a rare mineral, found in veins near or in igneous rocks. Associated minerals are bismuth, chalcopyrite, cassiterite, and wolframite. M.T. 1-A.

ORPIMENT

REALGAR

REALGAR
and
ORPIMENT

STIBNITE

BISMUTHINITE

5 cm

GALENA
PbS

Ore Mineral of Lead   **Isometric**

Perfect cubic cleavage and very brittle. Cubic, octahedral crystals, massive, granular. Lead gray, metallic luster and opaque. Lead gray streak. Specific gravity is 7.4-7.6 and hardness 2.5. BB fuses at 2, may decrepitate. In open tube yields sulphur fumes. Yields metallic globule in fusion with sodium carbonate in reducing flame. Globule malleable. Decomposed by HNO_3. Common mineral in veins, beds. Associate minerals pyrite, sphalerite, quartz, calcite, fluorite, barite and others. M.T. 1-A.

ARGENTITE (ACANTHITE)
Ag_2S

Ore of Silver   **Isometric**

Indistinct cleavage, uneven fracture. Cubic, octahedral, often branching, may be wiry, massive. Black gray, with tarnish to dull black. Metallic luster and opaque. Distinct shiny blackish gray streak. Hardness 2-2.5 and very sectile. Specific gravity 7.3. BB fuses at 1.5 producing silver globule in oxidizing flame. Globule dissolves in dilute HNO_3, add HCl to solution and yields white precipitate (AgCl). Uncommon, but in veins associated with other silver minerals, galena, sphalerite, tetrahedrite and cobalt and other minerals. M.T. 1-A.

CHALCOCITE
Cu_2S

Copper Ore Mineral   **Orthorhombic**

Indistinct prismatic, and conchoidal fracture. Crystals rare, usually massive, powdery. Dark lead gray and tarnishes black, blue, sometimes green. Metallic luster and is opaque. Streak a shiny gray-black. Hardness 2.5-3, somewhat sectile and specific gravity of 5.5-5.8. BB fuses at 2-2.5. Powdered and dissolves in HNO_3 solution turns blue when enough ammonium hydroxide is added. Open tube yields sulphur odor. Moistened with HCl and placed on Pt wire loop, gives blue-green flame (Cu). Common copper mineral of primary or secondary origin. Occurs in veins and in igneous rocks associated with enargite, chalcopyrite, bornite, galena, tetrahedrite, cuprite and other copper minerals. M.T. 1-A.

MOLYBDENITE
MoS_2

Ore Mineral of Molybdenum   **Hexagonal**

Basal perfect cleavage, flexible laminae, but not elastic. Hexagonal crystals, scaly, foliated, granular, massive. Bluish lead-gray, opaque, metallic luster and greenish gray streak. Hardness 1-1.5, somewhat sectile, greasy feel and specific gravity 4.7-4.8. Infusible, yields green flame. On charcoal BB mineral produces red coating near sample and yellow coating farther away. Latter to white when cool, and to blue when touched by reducing flame of blowpipe. Nitric acid decomposes mineral, leaves white residue. Associated with chalcopyrite, in pegmatites, veins and contact metasomatic deposits, also disseminated in granite like bodies. Scheelite, fluorite, wolframite, cassiterite, mineral associates.

GALENA

ARGENTITE

CHALCOCITE

5 cm

MOLYBDENITE

SPHALERITE   **Zinc Ore Mineral**   **Isometric**
ZnS
Contains Γe, Mn, Cd

Perfect dodecohedral cleavage (six directions), conchoidal fracture and brittle. Tetrahedrons, dodecohedrons, cubes, usually rounded crystals, massive, granular, some botryoidal. Yellow, brown, black, red, some white. Transparent to opaque and resinous to adamantine luster. Brown to yellow to white streak. Hardness 3.5-4 and specific gravity is 3.9-4.1. Fuses at 5 or infusible. Powder plus sodium carbonate heated BB on charcoal yields yellow coating (hot) and white (cold). Moisten with cobalt nitrate solution and heated BB strongly, turns green. Dissolve mineral powder in warm HCl and H_2S produced. Often fluoresces. Common and most important zinc mineral. Occurs in veins, replacement sulphide bodies associated with galena, pyrite, chalcopyrite and others.

PENTLANDITE   **Nickel Ore Mineral**   **Isometric**
(Fe, Ni) S

Octahedral cleavage, and uneven fracture, quite brittle. Metallic luster and distinctive light bronze-yellow color. Hardness is 3.5-4 and specific gravity 5.0. Light bronze-brown streak. Non-magnetic. May be intergrown with pyrrhotite and associated with other nickel minerals as millerite, niccolite, also with pyrite, chalcopyrite,. M.T. for Fe, Ni and S.

CINNABAR   **Mercury Ore Mineral**   **R-Hexagonal**
HgS

Perfect prismatic cleavage when visible, with uneven fracture. Rare crystals, rhombohedral, tabular, prismatic, acicular. Bright red to brownish red with adamantine luster when pure, dull when impure. Transparent to translucent. Gives red to brown streak. Hardness is 2-2.5, and specific gravity 8.1. BB volatile on charcoal; caution, toxic. In closed tube powdered cinnabar yields black sublimate, also when heated with dry sodium carbonate in closed tube produces metallic globules. Not a common mineral, occurs in veins, as disseminations in sedimentary and igneous rock, often with chalcedony. Associates are native mercury, realgar, stibnite, pyrite, marcasite. M.T. 1-B.

GREENOCKITE   **Cadmium Ore Mineral**   **Hexagonal**
CdS

Indistinct to distinct prismatic cleavage, and conchocidal fracture. Brittle. Rare, small crystals, encrustations. Orange, yellow, nearly transparent, adamantine to resinous luster. Orange streak, hardness is 3-3.5 and specific gravity 4.9-5.0. Infusible. Heat powdered mineral with sodium carbonate on charcoal BB and reddish brown coating of cadmium appears. In warm HCl yields H_2S. In closed tube when hot glows red, cooled returns to original color. Generally fluorescent. Often associated with sphalerite.

SPHALERITE

PENTLANDITE

CINNABAR

GREENOCKITE

5 cm

MILLERITE   Ore Mineral (minor) of Nickel   R-Hexagonal
NiS

Rhombohedral cleavage, rarely visible, uneven fracture and brittle. Needle-like crystals, radiating, crusts. Brass yellow, opaque, streak greenish black. Hardness is 3.0-3.5 and specific gravity 5.3-5.7. Metallic luster. Fuses at 1.5-2.0. On charcoal BB fuses and yields magnetic globule with odor of sulphur. Rare mineral. Mineral associates are calcite, fluorite, dolomite, hematite, siderite, pyrrhotite, and chalcopyrite. M.T. 1-A.

NICCOLITE   Nickel Ore Mineral   Hexagonal
NiAsS

No visible cleavage, uneven fracture, brittle. Usually massive, sometimes reniform, crystals rare. Pale copper red, opaque and metallic luster. Brownish black streak. Hardness is 5.0-5.5 and specific gravity 7.3-7.7. Fuses at 2.0 BB on charcoal to globule and yields garlic odor of arsenic and white coating at distance from sample. Powder dissolved in HNO_3 yields green solution. Rare mineral, occurs with cobalt, silver and other nickel minerals, pyrrhotite and chalcopyrite. M.T. 1-A.

PYRRHOTITE   Sometimes Ore Mineral of Nickel   Hexagonal
Fe_7S_8 When Disseminated Pentlandite in It

No cleavage, but basal parting, uneven fracture, brittle. Massive, granular, rare crystals, plates, tabular. Bronze, darkens with exposure to reddish bronze, opaque, metallic luster. Hardness is 3.5-4.5 and specific gravity 4.6. Fuses at 3.0. Magnetic in powder form. In open tube becomes very magnetic when heated. HCl decomposes it and yields H_2S. Common, occurs with pyrite, chalcopyrite, pentlandite, magnetite. M.T. 1-A.

COVELLITE   Ore Mineral   Hexagonal
CuS

Micaceous-like cleavage, uneven fracture. Tabular or platy, foliated and massive. Indigo blue to blue-black, purplish iridescent tarnish, metallic luster and opaque. Dark gray to black streak. Hardness is 1.5-2.0 with thin layers flexible. Specific gravity is 4.6-4.7. Burns with blue flame (Cu), sulphurous odor and fuses at 1.5 to 2.0. Moistened with water turns purple. Occurs in copper veins, disseminated copper bodies in igneous rock associated with bornite, chalcopyrite, chalcocite, enargite. Of probable secondary origin. M.T. 1-A.

NICCOLITE

SMALL CRYSTALS OF
MILLERITE

COVELLITE

5 cm

PYRRHOTITE

COVELLITE

BORNITE   Copper Ore Mineral   Isometric
Cu_5FeS_4

Poor cleavage and uneven fracture. Rare crystals, massive. Purplish to blue tarnish on weathered mineral, and bronze on fresh surface. Metallic luster, opaque and has grayish black streak. Hardness is 3 and specific gravity 4.9-5.4. BB fuses at 2.5. Is magnetic and brittle if heated BB in reducing flame on charcoal. If globule formed is moistened with HCl and heated BB it gives a blue flame. Mineral is soluble in HNO_3. Purplish tarnish a distinctive feature of mineral. Fairly widespread in copper-quartz veins, pegmatites and some igneous rocks. Associated minerals are chalcocite, chalcopyrite, covellite and other copper minerals. May be of primary or secondary origin.
M.T. 1-A

CHALCOPYRITE   Common Ore Mineral   Tetragonal
$CuFeS_2$ of Copper

Indistinct cleavage and uneven fracture. Tetrahedral crystals, usually massive. Brass yellow with common bronze or iridescent tarnish. Metallic luster, opaque with greenish black streak. Hardness 3.5-4, is brittle and has specific gravity of 4.1-4.3. BB fuses at 2 and on charcoal yields magnetic globule. Gives sulphur odor and decrepitates in closed tube. HNO_3 decomposes mineral. Associated with sphalerite, bornite, chalcocite, tetrahedrite, pyrrhotite, pyrite, galena and occurs in veins, contact metamorphic deposits, igneous rocks and some gneisses and schists. M.T. 1-A. Several solutions necessary. for M.T.

PYRITE   So-Called Fool's Gold   Isometric
FeS_2

Indistinct cleavage, when present cubic and octahedral, conchoidal to uneven fracture and brittle. Cubes, pyritohedrons, octahedrons, striated, massive, granular. Pale to medium brassy yellow. Opaque, metallic luster with black streak. Hardness is 6.0-6.5 and specific gravity 5.0. Fuses at 2.5-3.0. When heated yields magnetic globule and odor of sulphur dioxide. In closed tube gives yellow sublimate. If struck with steel, sparks. Common mineral associated with other sulfide minerals. Generally alters to limonite. M.T. 1-A.

BRAVOITE   Possible Nickel Ore Mineral   Isometric
$(Fe, Ni)S_2$

No cleavage, very brittle. Small grain crystals. Shows marked zoning under reflected light. Pale yellow, reddish tarnish. Opaque with metallic luster. Hardness varies with pyrite content from 6.0 for pyrite to 3.5 for nickel rich variety. Specific gravity 4.3-4.6, depending on nickel content. Reactions BB similar to pyrite. MT for iron, nickel and sulphur. A low temperature mineral. May develop from pentlandite, but may also occur in fissures, replacement masses and vugs.

BORNITE

CHALCOPYRITE

PYRITE

PYRITE

5 cm

COBALTITE — Cobalt Ore Mineral — Isometric
CoAsS

Perfect cubic cleavage and uneven fracture. Cubes, pyritohedrons, striated faces. Silver white, may have reddish tinge; metallic luster and opaque. Yields black streak. Hardness is 5.5 and is brittle; specific gravity is 6.0-6.3. BB fuses at 2 to 3. Heated on charcoal produces white coating away from mineral, has garlic odor (As) and gives magnetic globule. Heated in open tube gives sulphurous odor, and white arsenic ring. The borax bead with mineral turns blue (Co) when heated or cool. A rare mineral, occurs in veins with other cobalt and nickel minerals. M.T. 1A.

SMALTITE-CHLOANTHITE — Cobalt-Nickel-bearing Minerals — Isometric
CoAs – NiAs or RAs

Indistinct cleavage, uneven to granular fracture and brittle. Tin-white to steel-gray color and metallic luster. Hardness is 5.5-6.0 and specific gravity 5.7-6.8. May show iridescence or gray tarnish. Streak is grayish-black. Closed tube yields metallic sublimate of metallic arsenic, open tube a white sublimate of arsenic trioxide. BB yields arsenical odor, a white coating of arsenic trioxide on charcoal. Borax bead shows results for iron, cobalt and nickel. A vein mineral associated with cobalt and nickel.

MARCASITE — Orthorhombic
FeS_2

Indistinct to poor prismatic cleavage, uneven fracture, and brittle. Needle-like groups, radiating, botryoidal, stalactitic, crusts, granular, earthy. Pale brass yellow to silvery, brassy tarnish. Metallic luster, opaque, grayish-black streak. Hardness 6.0-6.5 and specific gravity 4.9. Fuses at 2.5-3.0 to magnetic globule. Yields SO_2 - yellow sublimate in closed tube. Common mineral in sedimentary rocks, in veins and replacements associated with galena, pyrite, chalcopyrite, calcite and dolomite. M.T. 1-A.

ARSENOPYRITE — Arsenic Ore — Orthorhombic
FeAsS

Distinct cleavage, uneven fracture. Prismatic, striated, twins, massive, granular. Silver-white to silvery-gray, opaque and with metallic luster. Streak dark grayish-blue. Hardness is 5.5-6.0, brittle and with specific gravity of 5.9-6.2. BB fuses at 2 to magnetic globule. Garlic-like odor when struck with hammer. Open tube test with powdered mineral yields white crystalline sublimate of As and SO_2 odor. Fumes turn moistened litmus paper red at end of open tube. Decomposed by HNO_3, yields sulphur. Common mineral, occurs in veins, pegmatites, contact metamorphic deposits and is associated with chalcopyrite, pyrite, sphalerite, cassiterite, and gold and silver minerals. M.T. 11-B.

MARCASITE

SMALTITE

ARSENOPYRITE

5 cm

COBALTITE

CALAVERITE   Ore Minerals of Gold and Silver   Monoclinic
$AuTe_2$
SYLVANITE
$(Au, Ag) Te_2$

Calaverite shows no cleavage, while sylvanite has one perfect cleavage, uneven fracture. Rare crystals, striated, granular, bladed, massive, branching. Both are silver-white to brass yellow with metallic luster and are opaque. Streak is gray. Hardness is 2.5 for calaverite and 1.5-2.0 for sylvanite. Both are brittle. Specific gravity 9.0-9.4 for calaverite, and 8.0-8.2 for sylvanite. BB fuses at 1, on charcoal block yields gold globule after long blowpiping. Because of silver content sylvanite yields silver-gold globule. Small quantity of powdered mineral in test tube with concentrated H_2SO_4, gently heated, turns solution reddish violet (Te). Rare minerals, occur in veins with gold, pyrite, quartz, fluorite, tetrahedrite and with each other. M.T. 1-A.

JAMESONITE   Lead Ore Mineral   Monoclinic
$Pb_4FeSb_6S_{14}$

Basal perfect cleavage, uneven fracture, quite brittle. Acicular crystals, Hair-like, fibrous, massive. Gray-black, lead gray, opaque and metallic luster. Streak gray black. Hardness 2-3, specific gravity 5.5-6.0. BB fuses at 1 and decrepitates. Almost disappears by volatilization BB, and yields yellow coating near sample on charcoal, white further away. When powdered and mixed with potassium iodide and sulphur, half and half, yields yellow coating BB on charcoal indicative of lead. Soluble in hot HCl. Uncommon, associated with galena, sphalerite, pyrite, tetrahedrite, bournonite, stibnite.

BOURNONITE   Ore Mineral of Lead,   Orthorhombic
$2PbS Cu_2S Sb_2S_3$ Copper, Antimony

Imperfect cleavage and uneven fracture. Thick tabular crystals, cross-like twins, granular. Gray-black to black, metallic luster and opaque. Streak gray-black to black. Hardness from 2.5-3.0, brittle and specific gravity of 5.7-5.9. BB fuses at 1. On charcoal BB gives white coating (Sb), and with continued blowpiping produces yellow coating (Pb). In closed tube intensely heated, gives brown coating (Sb). If powdered mineral is decomposed in HNO_3, solution turns blue when sufficient ammonium hydroxide is added (copper). Occurs in veins with sphalerite, galena, chalcopyrite, pyrite, tetrahedrite, stibnite and siderite. M.T. 1-A.

CALAVERITE

SYLVANITE

BOURNONITE

JAMESONITE

5 cm

PYRARGYRITE
Ag_3SbS_3

A Silver Ore Mineral
Dark Red

R-Hexagonal

Indistinct cleavage, conchoidal fracture and brittle. Deep red to dark gray. Ruby red for proustite. Reddish brown streak. Adamantine luster, nearly metallic when opaque, translucent to opaque. Hardness 2.0-2.5 and specific gravity is 5.5-5.9. Fuses at 1. BB fuses to globule with white coating close to sample (Sb), or at a distance for proustite (As). Uncommon, associated with argentite, tetrahedrite, galena, sphalerite, stephanite. M.T. 1-A.

PROUSTITE
Ag_3AsS_3

Ruby Silver Ore Mineral
Light Red

R-Hexagonal

Distinct cleavage, conchoidal to uneven fracture, quite brittle. Scarlet-vermillion color, transparent to translucent, especially on edges. Adamantine luster. Reddish streak. Hardness is 2.0-2.5 and specific gravity 5.6. BB fuses easily. In CT yields faint arsenic sublimate. On charcoal BB fuses, gives off sulphur and arsenic fumes. Under RF BB produces silver globule. M.T. 1-A. Associated with other silver minerals, generally in veins.

TETRAHEDRITE –
TENNANTITE
$(Cu,Fe)_{12}(SbAs)_4S_{13}$
with Ag, Pb, Hg, Zn substitutions

Copper Ore Mineral

Isometric

No visible cleavage, uneven fracture, brittle. Tetrahedral crystals, also massive and granular, tennantite dodecohedral. Dark gray to black, and opaque, shiny metallic luster. Black, brown to deep red streak. Hardness 3.0-4.0 and specific gravity 4.4-5.1. Fuses at 1.5. On charcoal BB yields white coating (As/Sb), and yellow coating if lead in composition, also produces green or blue flame (Cu) when heated, moistened with HCl and reheated. In veins, replacement bodies in limestone and associated with chalcopyrite, galena, sphalerite, pyrite, argentite and others. M.T. 1-A.

STEPHANITE
Ag_5SbS_4

A Silver Ore Mineral

Orthorhombic

Imperfect cleavage, subconchoidal to uneven fracture and brittle. Hardness is 2.0-2.5 and specific gravity 6.2-6.3. Massive to compact and disseminated. Metallic luster, iron black color and streak, opaque. Fuses in CT, gives faint sulphur fumes on long heating. Yields white coating on charcoal BB, also with extended heating a silver globule in charcoal. Soluble in dilute hot nitric acid. Associated with other silver minerals. M.T. 1-A.

TETRAHEDRITE

PYRARGYRITE

STEPHANITE

TENNANTITE

5 cm

ENARGITE   Copper Ore Mineral   **Orthorhombic**
Cu_3AsS_4

Perfect prismatic cleavage, with uneven fracture. Small crystals, tabular, prismatic, striated, also massive, granular. Grayish black, streak same. Metallic luster. Hardness is 3.0 and specific gravity 4.4-4.5. easily fused (1). BB on charcoal yields white coating, garlic odor (arsenic). In open tube produces white sublimate (As). BB on charcoal, moistened with HCl, gives green or blue copper flame. In veins and replacement bodies with chalcopyrite, bornite, tetrahedrite and others. M.T. 1-B then M.T. 1-A.

TENNANTITE-   Copper Ore Mineral   **Isometric**
TETRAHEDRITE
$(Cu, Fe)_{12}(SbAs)_4S_{13}$
with Ag, Pb, Hg, Zn substitutes

No visible cleavage, uneven fracture, brittle. Tetrahedral crystals, also massive and granular, tennantite, dodecohedral. Dark gray to black, and opaque, shiny metallic luster. Black, brown to deep red streak. Hardness 3-4 and specific gravity 4.4-5.1. Fuses at 1.5. On charcoal BB yields white coating (As/Sb), and yellow coating if lead in composition, also produces green or blue flame (Cu) when heated, moistened with HCl and reheated. In veins, replacement bodies in limestone and associated with chalcopyrite, galena, sphalerite, pyrite, argentite and others. M.T. 1-A.

STANNITE   Ore Mineral of Tin   **Tetragonal**
CuS FeS SnS

Indistinct cubic cleavage, uneven fracture and brittle. Steel gray to iron-black with blackish streak, and metallic luster. Hardness is 3.5 and specific gravity is 4.3-4.5. May tarnish bluish or be yellowish from presence of abundant chalcopyrite. Open tube yields sulphurous fumes, and a faint sublimate in CT. Globule on charcoal BB. In borax bead shows iron and copper reactions. Blue solution in nitric acid decomposition and separation of sulphur and tin dioxide. In tin-bearing veins associated with cassiterite, chalcopyrite, tetrahedrite and pyrite. M.T. 1-A

POLYBASITE   Brittle Ore Mineral   **Monoclinic**
$(Ag, Au)_{16}Sb_2S_{11}$ of Silver

Imperfect cleavage, uneven fracture and brittle. Massive and 6-sided plates with triangular marking, bevelled edges. Black with black streak, metallic luster and opaque. Very thin splinters of polybasite are dark red and translucent. Hardness is 2.0-3.0 and specific gravity 6.0-6.3. Fuses at 1. BB fuses to a globule on charcoal, yields white coating (Sb) close to sample and with sulphur odor. Rare silver mineral, in veins associated with argentite, galena, tetrahedrite, pyargyrite. M.T. 1-A.

ENARGITE

TENNANTITE

STANNITE

POLYBASITE

5 cm

SYLVITE   **Potash Salt**   **Isometric**
KCl

Perfect cubic cleavage, conchoidal fracture, brittle. Cubes with octahe-drons, massive, compact. Colorless to white, often tinted yellowish or reddish from impurities. Transparent to translucent and vitreous luster. Hardness 2.0 and specific gravity of 2.0. Slippery feel, dissolves in water, bitter salty taste. Fuses at 1.5 produces violet flame through blue glass (K). Occurs in salt deposits, rarer than halite.

CERARGYRITE   **Ore Mineral of Silver**   **Isometric**
AgCl

No cleavage, uneven fracture and very sectile. Crystals rare, usually cubes, massive, waxy. Colorless to greenish gray and pale gray. Darkens when exposed to light to violet brown. Adamantine to waxy luster and transparent to translucent. Streak is white to gray. Hardness 1-3 and specific gravity 5.5. BB fuses at 1, yields silver globule on charcoal. Rare secondary mineral in oxidized zone of ore deposits, native silver, cerussite associates. Dissolve in concentrated NH_4OH observe white AgCl crystals, cubes, crosses and stars. Insoluble in HNO_3.

FLUORITE   **An Ore Mineral**   **Isometric**
CaF_2

Perfect octahedral cleavage with smooth and flat fracture. Cubic crystals common, twined, massive, fine granular. Colorless, white, yellow, green, blue, purple, violet, pink, red, brown, sometimes with various shades in same sample. Vitreous luster and transparent to translucent. Hardness is 4.0 and mineral is brittle. Specific gravity 3.0-3.3. BB fuses at 3.0, has orange flame. May be fluorescent. If heated in darkened room continues to glow with removal of heat (phosphorescence). Common mineral in veins associated with quartz, calcite, galena, barite and others.

HALITE   **Rock Salt**   **Isometric**
NaCl

Perfect cubic cleavage, conchoidal fracture. Cubic crystals, massive, granular. Generally colorless to white, but tints of yellow, red, sometimes blue. Vitreous luster and transparent to translucent. Hardness is 2.5 and specific gravity 2.1-2.6. BB fuses at 1.5, readily soluble in water. Tastes salty. Gives strong yellow flame (Na) when fused on platinum wire loop. Occurs in salt flats, salt domes; associates are gypsum, calcite, sylvite (KCl), anhydrite and clay. Evaporite mineral.

SYLVITE

CERARGYRITE

HALITE

FLUORITE

5 cm

CARNALLITE Source Mineral for Potash Orthorhombic
(KMg) Cl$_2$6H$_2$O

No distinct cleavage, conchoidal fracture and brittle. Milkish white, also reddish. Greasy, shining luster, transparent to translucent. Hardness is 2.5 and specific gravity 1.6. Bitter taste and deliquescent. Occurs in saline deposits especially those with potash salts.

ATACAMITE Ore Mineral of Copper Orthorhombic
Cu Cl (OH)$_3$

One direction perfect cleavage and conchoidal fracture. Slender prisms, striated. Massive, granular, fibrous and sand-like. Bright green to blackish green, vitreous luster and transparent to translucent. Light green streak. Hardness is 3.0-3.5, brittle and specific gravity 3.8. BB fuses at 3.0-4.0, yields blue flame directly. Dissolves in dilute HNO$_3$. Add AgNO$_3$ and forms white precipitate (Cl). In closed tube produces water and turns litmus paper red. Non-effervescent in HCl. Uncommon copper mineral. Associated with other copper minerals, especially malachite and cuprite. M.T. 1-A.

CRYOLITE From the Greek Meaning Monoclinic
Na$_3$AlF$_6$ "Ice-Stone"

Parting, resembles cubic cleavage. Has uneven fracture. Cube-like crystals, massive. Mainly colorless to snow white, may be reddish to black. Transparent to translucent. Powdered mineral almost invisible in water with similar refractive index. Pearly luster. Hardness is 2.5, specific gravity 3.0. Easily fusible in candle flame. Soluble in sulphuric acid forming powerful hydrofluoric acid. A rare mineral from Greenland and Pikes Peak, U.S.A. from pegmatites.

CARNALLITE

CARNALLITE

CRYOLITE

ATACAMITE

5 cm

ILMENITE
FeTiO$_3$

A Titanium Ore Mineral   Hexagonal

No cleavage, conchoidal fracture, brittle. Thick, tabular, massive, compact, commonly twinned. Black opaque, metallic luster. Black to brownish red streak. Hardness is 5.0-6.0, specific gravity 4.3-5.0. Massive mineral, non-magnetic. Infusible. When powdered slightly magnetic. Powder fused with sodium carbonate on charcoal is magnetic. Latter dissolved in concentrated HCl and with tin added, solution boiled down, turns pale violet (titanium). Common mineral in veins, masses in igneous rocks. Also common in black sands. Associated with hematite and magnetite.

CHROMITE
Fc$_2$Cr$_2$O$_4$

Chromium Ore Mineral   Isometric

No cleavage and uneven fracture. Rare octahedral crystals, generally massive, granular. Black to brownish black, opaque and metallic luster. Brown streak and brittle. Hardness 5.5 and specific gravity 4.2-4.6. Powdered mineral slightly magnetic. Insoluble. BB and with borax bead and reheated bead has green color when cool (Cr). Occurs in lenses, pods, segregation layers in ultramafic igneous rocks, and in serpentine. Associates are serpentine, olivine, magnetite, corundum, uvarovite garnet. Often in black sands.

MAGNETITE
Fe$_3$O$_4$

An Iron Ore Mineral   Isometric

No cleavage, but octahedral parting, uneven fracture, quite brittle. Octahedra common, also granular, massive. Black, metallic luster, streak black, opaque. Hardness is 5.5-6.5 and specific gravity 5.2. Strongly magnetic. Infusible. Common mineral in igneous and metamorphic rocks, black sands. Also in large ore bodies. M.T. 11-A.

LODESTONE
Fe$_3$O$_4$

Strongly Polarized   Isometric
Magnetic Field

This variety of magnetite has the same properties described above, but shows distinctive polarity and radial alignment of the magnetic particles in a porcupine-like arrangement.

ILMENITE

CHROMITE

LODESTONE

MAGNETITE and APATITE

5 cm

CORUNDUM
Al_2O_3

Possible Gem-Mineral
Ruby = red, Sapphire = blue
Oriental Topaz = yellow, Oriental Amethyst = purple

Hexagonal

Pseudo-cleavage nearly rectangular, uneven to conchoidal fracture, and brittle. Blue (clear = sapphire), red (clear = ruby), yellow (clear = oriental topaz), brown and gray (corundum not transparent), and sometimes nearly white. Has adamantine to vitreous luster and is transparent to translucent. Hardness is 9.0 and specific gravity 3.9-4.1 Inert to acids. An accessory mineral in crystalline rocks such as granular limestone or dolomite, some gneiss, and micaceous and chlorite slate. Also rarely occurs in igneous rocks. May be present in placer deposits. Important localities are Burma, Sri Lanka, Thailand and Montana, U.S.A.

HEMATITE
Fe_2O_3

Most Important Iron Ore Mineral
Varieties: Specularite, Steel,
Earthy, Oolitic

R-Hexagonal

No distinct cleavage, but parting. Tabular, rhombohedral, columnar, massive, botryoidal and mammillary. Uneven or splintery fracture. Quite brittle. Steel gray to black, sometimes iridescent, red when earthy. Earthy to metallic luster. Good red streak. Hardness is 5.5-6.5, but to 1 in earthy red varieties. Specific gravity is 4.9-5.3. Infusible, BB in reducing flame becomes very magnetic. Slowly soluble in HCl. Accessory in igneous rocks, in veins with magnetite, contact metasomatic deposits, red cement in sandstones, other sedimentary rocks.

CUPRITE
Cu_2O

A Copper Ore Mineral

Isometric

Indistinct cleavage and uneven fracture. Octahedral, cubic, acicular, massive, granular. Reddish to blackish, subtransparent to subtranslucent on very thin edges. Brownish red streak, submetallic to adamantine luster. Hardness 3.5-4.0 and specific gravity 5.8-6.1. Resembles cinnabar in color and some hematite. Malachite, azurite, and chalcocite common associates. Fuses at 3.0. BB gives green flame. Fused with sodium carbonate yields copper globule on charcoal in reducing flame. With HCl and flame gives green tone. Soluble in concentrated HCl. M.T. 1-A.

ZINCITE
ZnO

Possible Ore Mineral of Zinc

Hexagonal

One direction perfect cleavage, uneven fracture and brittle. Rare crystals, platy, granular, massive. Deep red to orange-yellow. Translucent and subadamantine luster. Orange yellow streak. Hardness is 4.0 and specific gravity 5.4-5.7. Infusible. BB on charcoal powdered mineral mixed with sodium carbonate yields yellow coating (hot) and turns white (cold). Manganese substitutes for zinc in mineral and may produce Mn test. Soluble in acids. Rare, associated with calcite and franklinite.

CUPRITE and MINOR MALACHITE

SPECULAR HEMATITE

CORUNDUM

EARTHY HEMATITE

HEMATITE

5 cm

VUGGY HEMATITE

HEMATITE IN TACONITE

TENORITE   A Copper Mineral   Triclinic
CuO

Appearance of cleavage. Pseudo-monoclinic. Forms black scales with distinct metallic luster. May form earthy black masses with other copper minerals. Hardness is 3.0-4.0 and specific gravity 6.5. Good MT for copper. Lack of iron and other elements. Associated with cuprite, chrysocolla, and malachite in copper veins as oxidation product.

SPINEL   Transparent and Good Color   Isometric
$MgAl_2O_4$ Makes Gemstone
GAHNITE (Dark green)
$Zn\,Al_2O_4$

No cleavage, conchoidal fracture, brittle, variable colors, usually red (ruby spinel), black, green, blue, violet, brown. Octahedral crystals, some massive. Transparent to opaque and vitreous luster. Hardness is 8.0 and specific gravity 3.5-4.1. Insoluble and infusible. Powder dissolves completely in salt of phosphorus bead BB. This indicates absence of silica. Uncommon, in metamorphosed limestones, schists, also a placer mineral, accessory in basic igneous rocks. Associates are corundum, diopside, graphite, phlogopite, and others.

CHRYSOBERYL   Gemstone   Orthorhombic
$BeAl_2O_4$ Cat's Eye – dark olive green in daylight;
          red in artificial light (alexandrite)

Prismatic cleavage, but not prominent. Generally tabular to heart-shaped. Crystals often striated, twinned. Uneven to conchoidal fracture. Various greens, yellow, brown. Ceylon type has internal reflection, is green, and known as cat's eye. From USSR, a green variety which is red in artificial light, is called alexandrite. Vitreous luster and transparent to translucent. Hardness is 8.5 and specific gravity 3.5-3.8. BB infusible, also insoluble. Rare mineral, occurs in granite, pegmatite, mica schists and placer deposits.

CASSITERITE   Ore Mineral of Tin   Tetragonal
SnO_2

Indistinct cleavage and uneven fracture. Pyramidal or short prismatic, also reniform and fibrous, twins. Black to brown, rarely red or yellow, with adamantine to greasy luster and transparent to opaque. Streak pale brown to white. Hardness is 6.0-7.0 and specific gravity 6.8-7.1. BB infusible, also insoluble. If placed in dilute HCl with some zinc may slowly coat with zinc which brightens when rubbed. While rather widely distributed, rare as ore bodies of tin. Occurs in pegmatites and veins, also in placer deposits. Associated minerals include quartz, tourmaline, topaz, wolframite, scheelite, mica, apatite, fluorite and bismuthinite.

TENORITE

SPINEL (Gahnite)

CHRYSOBERYL

CASSITERITE

BROOKITE   Orthorhombic
TiO_2

Indistinct prismatic and basal cleavage with uneven to subconchoidal fracture. Brittle. Brown, yellowish, reddish to iron-black, all translucent and opaque. Uncolored streak to grayish or yellowish. Hardness 5.5-6.0, and specific gravity 3.8-4.1 Insoluble in acid. Tests similar to rutile and mineral association and occurrences like octahedrite.

RUTILE   Titanium Ore Mineral   Tetragonal
TiO_2

Distinct prismatic cleavage, uneven fracture, and brittle. Prismatic and pyramids, massive. Reddish brown, yellowish-red or black, transparent to opaque and adamantine to sub-metallic luster. Light brown streak. Hardness is 6.0-6.5 and specific gravity 4.2-4.3. Insoluble and infusible. Powder fuses with borax in Pt wire or on charcoal. Fine powder dissolves in HCl; with addition of tin and heating to boiling, the solution turns pale violet indicating Ti. Common mineral in small quantities. Scattered grains in igneous and metamorphic rocks, in veins, and in black sands. Common associates are quartz, feldspar, ilmenite, and hematite.

OCTAHEDRITE (Anatase)   Tetragonal
TiO_2

Perfect two-direction, not right angle cleavage. Subconchoidal fracture and brittle. Brown to indigo-blue and black, with adamantine to metallic adamantine luster. Transparent to almost opaque. Uncolored streak. Hardness is 5.5-6.0 and specific gravity 3.8-4.1. See rutile tests. A secondary mineral from titanium-bearing minerals. May occur in granite, quartz porphyry, mica and chlorite schists and gneiss. Associated minerals include hematite, apatite, titanite, rutile, brookite and quartz.

PYROLUSITE   A Manganese Ore Mineral   Tetragonal
MnO_2

Perfect prismatic, rarely observed, uneven to splintery fracture and brittle. Usually massive, often reniform coatings, dendrites, also fibrous, columns. Black to dark gray, opaque, metallic luster. Black streak. Hardness is 2.0-2.5 and specific gravity 4.7-4.9. Infusible. Fragment in borax bead BB, reheated in oxidizing flame gives violet tone when hot and reddish-violet cold – Mn test. Common as dendrites, also in irregular masses, nodules on sea floor. Associates are magnetite, psilomelane, hematite, barite, limonite. M.T. 1-B.

RUTILE PARAMORPH
after Brookite

ANATASE

PYROLUSITE

5 cm

PSILOMELANE   Manganese Ore Mineral   Orthorhombic
Variable Manganese Oxides

No cleavage, uneven to conchoidal fracture, and brittle. Massive, botryoidal, stalactic. Dark gray to black, opaque and submetallic to dull luster. Streak brownish-black to black. Hardness is 5.0-6.0 and specific gravity 3.3-4.7. Infusible, Oxidizing flame BB bead is violet when hot (manganese) and reddish-violet when cold. Soluble in HCl, yields water in closed tube. Uncommon mineral associated with pyrolusite, manganite, calcite, and hematite. M.T. 1-B.

DIASPORE   Important Mineral of Bauxite   Orthorhombic
$Al_2O_3 \cdot H_2O$

Occurs in microscopic-sized crystals with pisolitic aggregates. Platy, tabular, also massive. Has one good cleavage. Somewhat translucent, with pearly luster on cleavage face, other vitreous. White and gray to brown, pink. Specific gravity is 3.0-3.1 and hardness 6.5-7.0. Associates are boehmite and gibbsite; also with corundum and emery.

GOETHITE   Ore Mineral of Iron   Orthorhombic
$FeO(OH)$

One perfect cleavage, uneven fracture, and brittle. Rare crystals, platy, prismatic, bladed, botryoidal, mammillary, fibrous. Usually very dark brown to black. Earthy variety is ocherous yellow-brown. Subtranslucent, thin fragments transparent; some approach metallic luster. Brownish-yellow streak. Hardness is 5.0-5.5 and specific gravity 3.3-4.3. Soluble in HCl. BB difficultly fusible at 5.0-5.5. Magnetic after heat BB in reducing flame. Yields water in closed tube. Widely distributed, forms from weathering of iron-bearing minerals, e.g., siderite, pyrite, magnetite. M.T. II-A.

MANGANITE   A Manganese Ore Mineral   Orthorhombic
$MnO(OH)$

One perfect cleavage, uneven fracture and brittle. Generally prismatic, vertically striated, bundle form, columnar, stalactic, radiating, fibrous. Dark grey to black, submetallic luster and dark brown streak. Hardness 4.0 and specific gravity is 4.2-4.4. Infusible. Powdered mineral in small quantity fused in borax bead BB and reheated in oxidizing flame shows violet (hot) and reddish violet (cold) indicative of manganese. Water in closed tube, and soluble in HCl. Uncommon secondary mineral in veins, associated with barite, siderite, calcite, limonite, and psilomelane. M.T. 1-B.

PSILOMELANE

DIASPORE

MANGANITE

5 cm

GOETHITE

LIMONITE   Iron Ore Mineral   Not Crystallized
2 Fe O 3H O

No distinct cleavage, stalactic, botryoldal, or mammillary forms common. Internal fibrous or sub-fibrous structure; earthy, concretionary and also massive. Brown shades, usually dull, may have nearly black exterior. Common rusty appearance. Non-metallic to sub-metallic luster, streak yellowish brown. Hardness is 5-5.5 and specific gravity 3.6-4.0. An alteration product of other iron-bearing minerals. BB in salt of phosphorus bead yields siliceous skeleton. Water in tube tests, Fe in MT. A common mineral.

WAD   Manganese Ore Mineral   Amorphous
Mn oxide

Soft, soils fingers, earthy or compact. Dull black, bluish, brownish black. Hardness very soft to 6. Specific gravity 3.0-4.3. Associated with other manganese-bearing oxide minerals.

BAUXITE   Ore of Aluminum   Mixture
$Al_2O_3 2H_2O$

Imperfect or non-existent cleavage, earthy fracture. Small rounded nodules in claylike matrix. Massive, earthy. Generally white, gray yellow, brown and red. Dull or earthy luster, opaque. Streak white cream, yellow, brown or red. Hardness 1-3, quite crumbly and brittle with specific gravity of 2.0-2.6. BB infusible, also insoluble. Moisten with cobalt nitrate and BB turns blue (Al). Yields water in closed tube. Adheres to tongue, clayey odor, non-plastic with water. Common mineral, generally in tropical climates from weathering of aluminum rich rocks.

BRUCITE   Magnesium Ore Mineral   Hexagonal
$Mg(OH)_2$

Mica-like, platy cleavage, basal perfect, flexible but not elastic layers. Broad tabular crystals, also fibrous, massive, foliated. White, light gray, pale green with vitreous to waxy luster, somewhat pearly on cleavage face. Transparent to opaque. Hardness is 2-2.5 and mineral is sectile with a specific gravity of 2.4. Usually fluorescent and is soluble in acids. BB infusible, but glows. In closed tube yields water. Not uncommon, occurs in veins, metamorphosed dolomites, associated with talc, calcite, chlorite, magnesite.

GIBBSITE   Aluminum Ore Mineral   Monoclinic
$Al_2O_3 3H_2O$

Basal cleavage, tough. White, grayish greenish or reddish white. Luster pearly on basal cleavage, vitreous on other faces. Translucent, hardness is 2.5-3.5 and specific gravity 2.3-2.4. Clayey odor when breathed on. Gives water in CT and becomes white. Infusible BB. Soluble in concentrated sulphuric acid, ignited with cobalt solution shows deep blue color. It is a secondary mineral from the alteration of aluminum-bearing minerals. Associated with bauxite and in laterites.

LIMONITE

WAD

BAUXITE

BRUCITE

5 cm

GIBBSITE

CALCITE
CaCO₃

Perfect rhombohedral cleavage and conchoidal fracture. Varied crystals common, tabular, prismatic, acute, dog-tooth like. Usually colorless or white, often light tones of gray, yellow, green, blue, red. Has vitreous to pearly luster and is transparent to opaque. Hardness 2.5-3.0, and specific gravity of 2.7. Effervesces in HCl. May fluoresce. An abundant mineral in rocks, veins, spring deposits, cave deposits, and shells of sea animals. Makes up limestone rock.

Calcite, the CaCO₃ mineral, is the dominant member of the **Calcite Group**. The main minerals of this group include dolomite, ankerite, magnesite, siderite, rhodochrosite, smithsonite and the rare sphaerocobaltite. The group minerals crystallize in the rhombohedral class of the hexagonal system and have characteristic rhombohedral cleavage with slight variations in the angle between 75° and 105° for calcite and 73° and 107° for siderite. The latter has almost identical internal structure to rhodochrosite, the MnCO₃ mineral. The iron and manganese molecules are almost completely miscible.

Aragonite, also of CaCO₃ composition, differs in its crystal system and is the dominant mineral of the **Aragonite Group**. The latter includes witherite, strontianite and cerussite of this book. Each crystallizes in the orthorhombic system. Prismatic cleavage shows angles of small variation from 60° and 120°. X-ray data suggest an internal structure nearly hexagonal in character, supporting the close relationship of the two groups.

CALCITE VARIETIES

5 cm

DOLOMITE   Rock and Mineral   R-Hexagonal
$CaMg(CO_3)_2$

Good three-direction cleavage, not right angles, conchoidal fracture. Rhombohedral, often curved, massive, granular. Generally white, gray or pink, can be colorless. Vitreous luster to pearly and dull, transparent to translucent. Quite brittle, hardness is 3.5-4.0 and specific gravity 2.8-2.9. Effervesces in hot HCl only. BB infusible. Common associate with calcite, may occur in veins and as alteration product of limestone.

MAGNESITE   A Carbonate Mineral   R-Hexagonal
$MgCO_3$

Rhombohedral cleavage, conchoidal fracture and brittle. Crystals rare, mainly massive, granular, or earthy. White, light gray, yellowish brown, transparent to opaque, vitreous to dull luster. Hardness 3.5-5.0 and specific gravity 3.0-3.2. Powder effervesces in hot, dilute HCl, not in cold HCl. Often in veins in serpentine; also in granular form in sedimentary rocks.

SIDERITE   Iron Ore Mineral   R-Hexagonal
$FeCO_3$

Rhombohedral perfect cleavage, conchoidal to uneven fracture and brittle. Rhombohedral crystals, curved faces, massive, granular, globular, botryoidal, fibrous, earthy. Light brown to dark brown, sometimes grayish and can be transparent to translucent, and vitreous to pearly luster. Hardness is 3.5-4.0 and specific gravity 3.8-3.9. Dissolves and effervesces in hot dilute HCl. Fuses at 4.5, BB becomes black and magnetic. In closed tube decrepitates, blackens and is magnetic. Common mineral, occurs in limestone, clay, shale, coal, forms ironstone. Also in veins with galena, pyrite, chalcopyrite, and tetrahedrite.

RHODOCHROSITE   Manganese Ore Mineral   R-Hexagonal
$MnCO_3$

Perfect rhombohedral cleavage, uneven fracture and brittle. Rhombohedral crystals, curved faces, granular, massive, botryoidal, crusts. Rose-pink, occasionally light gray to brown. Sometimes stained black. Transparent to translucent, vitreous luster. White streak. Hardness is 3.5-4.5 and specific gravity 3.4-3.6. Infusible. Darkens and decrepitates BB. Effervesces in powder form in hot HCl. Powder in borax bead BB in oxidizing flame turns bead violet (hot) and reddish-violet (cold), indicative of Mn. Uncommon, with manganese minerals and lead, silver, copper minerals.

DOLOMITE AND SMALL
CRYSTALS OF CHALCOPYRITE

DOLOMITE

MAGNESITE

RHODOCHROSITE

SIDERITE

5 cm

SMITHSONITE
$ZnCO_3$

Zinc Ore Mineral
Gem Mineral

R-Hexagonal

Rare cleavage, but rhombohedral perfect, uneven to almost conchoidal fracture and brittle. White, grayish, greenish, blue, brown, sometimes pink or yellow. Rare crystals, rhombohedral with curved faces, botryoidal reniform, stalactic, encrusting. Translucent and vitreous luster. White streak. Hardness is 4.0-5.5 and specific gravity 4.3-4.4. Effervesces in cold HCl. Infusible. Powder mixed with sodium carbonate and heated BB on charcoal yields yellow coating hot, white when cold. Add cobalt nitrate solution to coating, heat strongly BB and will turn green (Zn). M.T. 1-A.

ARAGONITE
$CaCO_3$

Orthorhombic

One direction of cleavage poor, uneven fracture. Acicular, tabular, prismatic, fibrous, and encrustations. Colorless, white, light gray, yellow, with vitreous luster, and transparent to translucent. Hardness is 3.5-4.0, brittle with specific gravity of 2.9-3.0. BB infusible, but decrepitates. In HCl effervesces, in closed tube turns to powder. Moisten with HCl, heat, and gives red flame. Fluorescence generally orange or green. Very common mineral, associated with gypsum, siderite, celestite, sulphur, limonite, and others.

ANKERITE
$CaCO_3 (Mg, Fe, Mn) CO_3$

Member of Calcite Group

R-Hexagonal

Distinct rhombohedral cleavage and brittle. Crystals, massive, granular. White, grayish and reddish tones. Transparent to translucent with vitreous luster. Hardness is 3.5, specific gravity 3.0-3.1. Effervesces in warm HCl. MT for iron and manganese. Associated with iron ores, also in veins.

BASTNAESITE
$(RF)CO_3$

Cerium Metal Source

Hexagonal

Basal parting, prismatic crystals. Translucent to opaque. Wax yellow to reddish born. Non-metallic luster. Hardness is 4.5 and specific gravity 4.9. BB infusible. In sands, also in carbonatites associated with barite in large deposit in California.

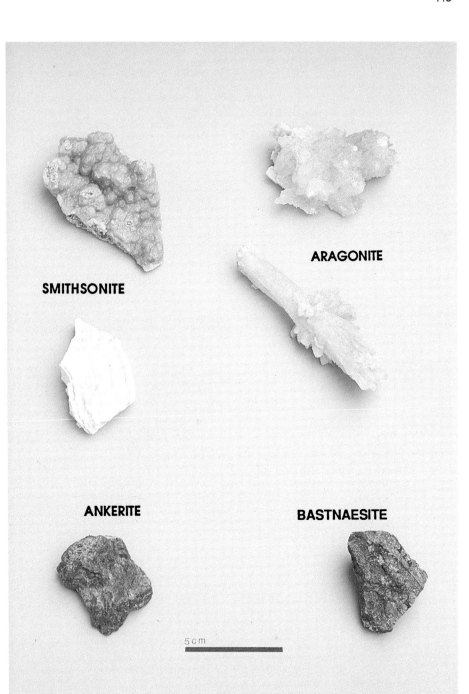

SMITHSONITE

ARAGONITE

ANKERITE

BASTNAESITE

5 cm

CERUSSITE — Lead Ore Mineral — Orthorhombic
$PbCO_3$

Prismatic cleavage, conchoidal fracture. Prismatic, tabular, acicular, granular, massive, often twinned. Colorless, white, light gray with adamantine luster and transparent to translucent. Hardness 3.0-3.5 and very brittle. Specific gravity of 6.5-6.6. White streak. BB fuses at 1.5. Lead globule forms when mineral is mixed with sodium carbonate, heated on charcoal BB, also yellow to white coating. Effervesces in warm dilute HNO_3. May fluoresce green, blue or yellow. Common secondary lead mineral and is associated with galena, anglesite, limonite, pyromorphite. Occurs in veins and bedded replacement bodies. M.T. 1-A.

WITHERITE — Barium Source Mineral — Orthorhombic
$BaCO_3$

One direction good cleavage, uneven fracture and brittle. Pseudo-hexagonal pyramids, columnar, globular, botryoidal. Colorless, white and grayish. Translucent with vitreous luster. Hardness is 3.0-3.5 and specific gravity 4.3. Fuses 2.5-3.0 and effervesces in dilute HCl. Powder moistened with HCl yields yellowish green flame BB. May fluoresce as mineral. Rare, with galena in veins.

STRONTIANITE — Strontium Source Mineral — Orthorhombic
$SrCO_3$

Distinct prismatic cleavage, uneven fracture, brittle. Prismatic, acicular, fibrous, granular, massive. White, colorless, light gray, yellow, light green. Transparent to translucent with vitreous luster. Hardness is 3.5-4.0. and specific gravity 3.7. Infusible, but swells, produces red flame (Sr). Effervesces and dissolves in HCl. Platinum loop immersed in latter solution gives red flame over bunsen burner. Possible fluorescence. Uncommon mineral. Occurs in veins, limestones, igneous rocks.

TRONA — Ore Mineral of Soda Ash — Monoclinic
$Na_3H(CO_3)_2 2H_2O$

Perfect prismatic cleavage, uneven to subconchoidal fracture. Plates, fibrous, columnar, massive. Gray or yellowish-white, vitreous luster and translucent. Hardness 2.5-3.0 and specific gravity 2.1. Alkaline taste, soluble in water, yields water and carbon dioxide in closed tube. Intense yellow flame BB and readily fusible at 1.5. Effervesces in acid. Occurs in saline lake deposits, as efflorescences and in thick, extensive beds as near Green River, Wyoming.

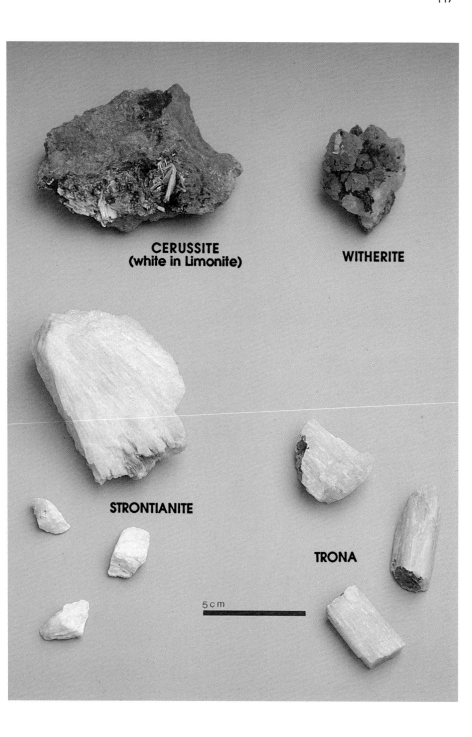

CERUSSITE
(white in Limonite)

WITHERITE

STRONTIANITE

TRONA

5 cm

MALACHITE   Copper Ore Mineral   Monoclinic
$Cu_2CO_3(OH)_2$

Cleavage one direction perfect, but rarely visible, uneven fracture, and brittle. Bright green, black surface often visible. Translucent to opaque and vitreous (crystals), silky (fibrous), dull (earthy). White to pale green streak. Hardness is 3.5-4.0 and specific gravity 3.9-4.0. Fuses 2-3 with green (copper) flame. Effervesces in HCl and yields water in closed tube. Common copper mineral associated with azurite, cuprite, native copper. M.T. 1-A.

AZURITE   Ore Mineral of Copper   Monoclinic
$Cu_3(CO_3)_2(OH)_2$

One good cleavage, two poor and conchoidal fracture. Crystal groups, botryoidal, massive, fibrous, radiating, sometimes earthy. Azure blue with blue streak. Vitreous luster and is translucent to opaque. Pale blue streak. Hardness from 3.5-4.0 and brittle, with 3.8 specific gravity. BB fuses at 3, produces green flame (Cu), and effervesces in HCl. Yields water in closed tube. Copper globule results when powdered mineral mixed with equal parts of sodium carbonate and charcoal are intensely heated in reducing flame on charcoal block. Common secondary copper mineral associated with malachite in oxidized zone.

AURICHALCITE   Monoclinic
$2(Zn,Cu)CO_3\text{-}3(Zn,Cu)(OH)_2$

Perfect cleavage, brittle. Pale green to sky-blue. Transparent to translucent and with pearly luster. White streak. Hardness is 2.0-2.5 and specific gravity 3.6. Effervesces in acid. MT for copper. A secondary mineral associated with malachite; relatively rare.

MALACHITE

AZURITE

AURICHALCITE

NITER   Known as Saltpeter   Orthorhombic
KNO_3

Cleavage not distinct, uneven fracture and brittle. Needle-like crystal groups, thin crusts and coatings. Colorless to white and gray, transparent to translucent with vitreous luster. Hardness is 2.0 and specific gravity 2.1. Fuses at 1 and yields violet potassium flame viewed through blue-glass filter. Soluble in water. Salty taste. Occurs in arid regions and caves.

SODA NITER   Chile Saltpeter   R-Hexagonal
$NaNO_3$

Rhombohedral for crystals, conchoidal fracture, but rarely observable, and somewhat sectile. Massive, granular and crusts. Colorless to white. Sometimes tinted gray, yellow or reddish brown from impurities. Transparent to translucent and vitreous luster. Hardness 1.5-2.0 and specific gravity 2.2. Fuses at 1, gives strong yellow flame. Is soluble in water. Has cooling taste and absorbs moisture from air. Is in arid regions.

BORACITE   Borate Mineral   Pseudo-Isometric
$5Mg_6Cl_2B_{14}O_{26}$ Orthorhombic

Trace of cleavage, fracture conchoidal and uneven brittle. White to gray, yellow and green. White streak. Subtransparent to translucent. Vitreous to somewhat adamantine luster. Hardness is 7 in crystals and specific gravity is 2.9. In massive form yields water in CT. Fuses BB at 2 swelling to white pearl, flame colored green. Soluble in HCl.

TINCALCONITE   Borate Mineral
$Na_2O\ 2B_2O_3\ 5H_2O$

Former name "Mohavite". Generally lusterless white small crystals or coatings on borax and kernite. Specific gravity of 1.88, alteration product of borax and kernite; contains up to 32% water.

BORAX-KERNITE   Ore Mineral of Borax   Monoclinic
$Na_2B_4O_7\ 10H_2O$

One good cleavage, but rarely seen. Often large prismatic, also massive cellular, and as crusts. Has conchoidal fracture. Colorless, white, grayish with vitreous luster and is translucent to opaque. Hardness is 2.0-2.5, brittle, and specific gravity is 1.7. BB swells and fuses at 1.0-1.5 to clear globule with strong yellow flame. Has sweet alkaline taste. Soluble in cold water. In closed tube yields much water. Fairly common mineral in arid regions, lake deposits and coatings on soil. Associated minerals are halite, colemanite, ulexite, and gypsum. Kernite resembles borax, but easily cleaves into narrow splinters. Known only in Kern County, California, where it occurs with borax and ulexite.

121

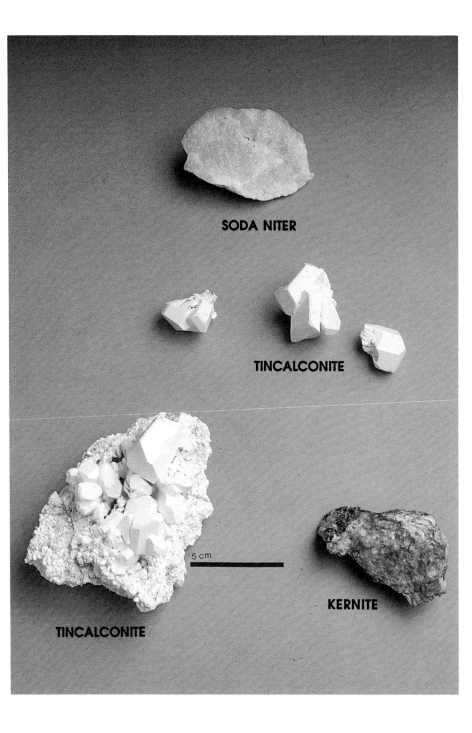

SODA NITER

TINCALCONITE

5 cm

TINCALCONITE

KERNITE

ULEXITE　　　　　Borax-bearing Mineral　　　　　Monoclinic
$NaCaB_5O_9\,8H_2O$

Rounded masses, white, loose texture, fibrous, acicular. Hardness is 1.0 and specific gravity 1.7. Silky luster. Tasteless, not soluble in cold water, slightly in hot water. Latter alkaline reaction. CT and OT yields water. Fuses at 1 BB with swelling to clear blebby glass. Yields deep yellow flame; add sulphuric acid and momentarily changes to deep green.

COLEMANITE　　　　　A Borate Mineral　　　　　Monoclinic
$Ca_2B_6O_{11}\,5H_2O$

One direction perfect cleavage, uneven fracture. Crystals variable, usually short prismatic, also massive, granular. Colorless to white with vitreous luster and is transparent to translucent. Hardness is 4.0-4.5, brittle and has a specific gravity of 2.4. BB fuses at 1.5, decrepitates violently, produces green (B) flame. May be fluorescent. Will dissolve in hot HCl and when cooled precipitates boric acid. Occurs in association with borax, ulexite in sedimentary rocks. Usually of lacrustine origin.

BARITE　　　　　Ore Mineral　　　　　Orthorhombic
$BaSO_4$

Two direction perfect cleavage, right angles, uneven fracture. Tabular, often divergent, barite roses, granular, earthy. White or colorless, gray, yellow, blue, red, or brown tints. Vitreous luster and sometimes pearly on cleavage face. Transparent to opaque. Hardness is 2.5-3.5 is brittle and has a specific gravity of 4.5, diagnostic for a non-metallic mineral. BB decrepitates, whitens, and fuses at 3-4. Yields yellowish-green flame (Ba). Fused mineral turns moistened red litmus paper blue. Insoluble in acids. A common mineral in veins, in beds, residual masses in clay.

CELESTITE　　　　　Strontium Source Mineral　　　　　Orthorhombic
$SrSO_4$

Basal perfect and prismatic cleavage at right angles, and uneven fracture. Acicular, also fibrous, reniform or granular. Colorless, white, bluish or reddish with vitreous to pearly luster and transparent to translucent. Hardness is 3.0-3.5 and specific gravity 3.9-4.0. BB fuses at 3.5-4.0, gives red strontium flame, usually decrepitates. Insoluble. May fluoresce. Rare mineral, occurs in sedimentary rocks, also in veins. Mineral associations include dolomite, gypsum, halite, calcite, fluorite, barite.

124

ANGLESITE · Ore of Lead · Orthorhombic
$PbSO_4$

Prismatic cleavage, but absent in massive or earthy form; conchoidal fracture. Tabular, prismatic, pyramidal, also massive, granular. Generally white or grayish, with adamantine to resinous luster. Transparent to opaque. Hardness is 3.0 and mineral is very brittle; specific gravity 6.3–6.4. BB fuses at 2.0 and decrepitates. Mix powdered mineral with sodium carbonate and BB on charcoal block, and in reducing flame, yields lead globule. Dissolves slowly in cold HNO_3 acid; fluorescence common. Mineral occurs in oxidized zones of lead deposits; associates are cerussite, wulfenite, smithsonite, hemimorphite, and pyromorphite. M.T. 1-A.

ANHYDRITE · Orthorhombic
$CaSO_4$

Three directional cleavage at right angles, fracture is uneven to splintery. Rare crystals, massive, granular, fibrous. Colorless to white, but may be tinted blue, red, or gray. Vitreous luster, somewhat pearly on cleavage, may be translucent to opaque. Hardness is 3.5 and specific gravity 2.9–3.0. BB fuses at 3, decrepitates slightly. Dissolves in hot HCl and gives white precipitate when $BaCl_2$ is added, proving sulphate content. Moisten with HCl and mineral gives orange flame (Ca). Can be fluorescent. Harder than gypsum, and a common associate with it, salt and limestone. Occurs also in veins with metallic minerals, also in openings in basalt.

GYPSUM · An Important Industrial Mineral When Pure and in Quantity · Monoclinic
$CaSO_4 2H_2O$

Alabaster – massive
Selenite – colorless – transparent
Satinspar – fibrous, silky

One direction perfect cleavage, conchoidal and uneven fracture. Tabular crystals, also fibrous, massive, granular. Colorless, white, pale gray, sometimes tints of yellow or reddish. Transparent to opaque. Streak white. Vitreous luster, somewhat pearly parallel to cleavage. Hardness is 2.0 and specific gravity 2.3. BB fuses at 3.0. Soluble in hot HCl, add barium chloride and yields white precipitate of $BaSO_4$. In closed tube produces water. Evaporite mineral most often, associated with salt, anhydrite, limestone, clays.

BROCHANTITE · Copper Mineral · Orthorhombic
$4CuO SO_3 3H_2O$

Perfect cleavage one direction, less so another, uneven fracture. Emerald to blackish green, vitreous luster, transparent to translucent. Hardness is 3.5–4.0 and specific gravity 3.9. Streak is green. In CT yields water, and at high temperatures sulphuric acid and turns black. Fuses BB and on charcoal reduces to metallic copper. An oxidation product of copper minerals, in oxidized zone.

ANGLESITE

ANHYDRITE

(Alabaster)

(Satinspar)

GYPSUM

(Selenite)

5 cm

BROCHANTITE

ALUNITE
$KAl_3(SO_4)_2(OH)_6$

Possible Source of Alum, Potassium and Aluminum

R-Hexagonal

Imperfect cleavage and uneven or splintery fracture, also conchoidal. Rare crystals, rhombohedral, pseudo-cubic, often massive. Generally white, sometimes pale gray, and crystalline variety pinkish. Hardness 3.5-4.0, quite brittle; specific gravity 2.6-2.8. Vitreous to pearly luster, commonly transparent to translucent. Usually massive (replacement type), though rare crystals are usually rhombohedrons in hexagonal system. BB yields violet flame (potassium), also decrepitates; in closed tube produces acid water; is soluble in H_2SO_4. An uncommon mineral, generally associated with alteration zones related to hydrothermal activity in shallow seated environments.

GLAUBERITE
$Na_2Ca(SO_4)_2$

Evaporite Mineral

Monoclinic

Perfect basal cleavage and conchoidal fracture and brittle. Prismatic or tabular crystals. Vitreous luster, transparent to translucent. White streak. Flesh pink to brick red. Hardness is 2.5-3.0 and specific gravity 2.7-2.8. BB fuses at 1.5 decrepitates and has yellow flame (Na). Powder dissolves in HCl, add $BaCl_2$ and white precipitate of $BaSO_4$ forms. Partially and slowly dissolves in water. Bitter, salty taste. Uncommon, in salt lake deposits associated with halite, thenardite, and other salts.

CHALCANTHITE
$CuSO_4\,5H_2O$

Possible Ore Mineral of Copper

Triclinic

Cleavage, 3-directions, poor. Stalactitic, fibrous, veinlets, triclinic crystals. Blue to greenish-blue crystals. Massive, also fibrous, conchoidal fracture. Vitreous luster, somewhat translucent. White streak. Hardness is 2.5 and specific gravity 2.1-2.3. Readily soluble in water and has nauseating taste. In dry air turns to greenish-white powder from dehydration. BB fuses at 3 and with soda on charcoal yields metallic copper. Commonly associated with other copper sulfate minerals and with gypsum. Forms stalactitic coatings in copper mine workings. Of widespread occurrence in desert southwest of U.S. in areas of copper mineralization.

EPSOMITE
$MgSO_4\,7H_2O$

An Evaporite Mineral

Orthorhombic

One direction perfect cleavage and conchoidal fracture, fairly brittle. Botryoidal, fibrous crusts, crystals rare. White with vitreous to earthy luster, and transparent to translucent. Hardness is 2.0-2.5 and specific gravity 1.7. BB fuses at 1. Soluble in water, bitter taste (epsom salts). In closed tube test mineral liquifies and gives red litmus (acid) test. Uncommon, though occurs in caves or mine openings as encrustations.

ALUNITE
(Fine-grained)

GLAUBERITE

ALUNITE (Crystalline)

CHALCANTHITE and
IGNEOUS ROCK

5 cm

CHALCANTHITE

MONAZITE
(Ce LaYTh)PO$_4$

Rare Earth Thorium
Ore Mineral

Monoclinic

Distinct cleavage, pinacoidal, conchoidal to uneven fracture, brittle. Flattened crystals, wedge-shaped, prismatic, tabular. Reddish brown, clove brown, sometimes greenish. Translucent to opaque. Resinous to waxy luster. Hardness is 5.0-5.5 and specific gravity 4.9-5.3. Infusible. BB turns gray on charcoal. Insoluble in HCl. Sulphuric acid on mineral BB yields brief blue flame of PO$_4$. Radioactive. Rare mineral, accessory in pegmatites, granite, some gneisses, heavy mineral in stream sands with magnetite, rutile, zircon, ilmenite, and others.

APATITE
Ca$_5$(F$_3$Cl)(PO$_4$)$_3$

Gemstone when Transparent,
of Good Color and Crystal Form

Hexagonal

Imperfect cleavage and conchoidal and uneven fracture. Crystals, usually hexagonal, prismatic, pyramidal, massive, granular. Generally green or brown, but may be white, blue, violet or colorless, mottling common. Vitreous luster. Transparent to opaque. Hardness is 5, brittle, with 3.1-3.2 specific gravity. BB fuses difficultly at 5, yields reddish-yellow flame. Moisten mineral with H$_2$SO$_4$ over heat BB to produce bluish-green flame (PO$_4$). Soluble in acids; in HCl add dilute H$_2$SO$_4$ drops and white precipitate indicates Ca. Usually fluorescent. Common mineral in sedimentary beds (phosphates), pegmatites, veins, metamorphosed limestones, in magnetite veins, and minute crystals in igneous rocks.

PYROMORPHITE
Pb$_5$Cl(PO$_4$)$_3$

Rare Ore Mineral
of Lead

Hexagonal

Indistinct cleavage, uneven fracture and brittle. Simple prismatic, rounded and barrel forms. Shades of green, yellow, brown for pyromorphite and white, pale yellow to yellow brown (mimetite). Resinous luster and translucent. Hardness is 3.5-4.0 and specific gravity 6.5-7.1. Fuses at 2. BB fuses to globules. Dissolve in nitric acid and add warm ammonium molybdate and yellow precipitate indicates phosphate. Garlic odor suggests As of mimetite BB on charcoal. Rare mineral in oxidized zone of lead deposits. M.T. 1-A.

MIMETITE
(PbCl)Pb$_4$(AsO$_4$)$_3$

A Lead Arsenate

Hexagonal

Imperfect cleavage, uneven fracture and brittle. Pale yellow to brown, orange, yellow-white and colorless. White streak. Resinous luster, subtransparent to translucent. Hardness is 3.5 and specific gravity 7.0-7.2. Yields white sublimate of lead chloride in closed tube. Fuses at 1 BB. In reduced flame BB on charcoal gives arsenical odor, reduces to metallic lead. Soluble in HNO$_3$. With powdery charcoal in CT yields black and gray, silvery-gray sublimates of As. M.T. 1-A.

MONAZITE

APATITE

PYROMORPHITE

MIMETITE

5 cm

VANADINITE
Ore Mineral of Vanadium
Hexagonal

$Pb_5Cl(VO_4)_3$

No visible cleavage, uneven fracture and brittle. Sharp and prismatic, may be hollow crystals. Ruby red, orange-red, yellow, reddish brown. Translucent to opaque, resinous luster. White to yellowish streak. Hardness is 3 and specific gravity 6.7-7.1. Fuses at 1.0-1.5. Powder in sodium carbonate bead BB on charcoal yields lead globule. Decrepitates and has faint white sublimate of lead chloride in closed tube. Powder in salt of phosphorous bead produces yellow color in oxidizing flame and green color in reducing flame BB. Rare, in oxidized zone of sulphide ore deposits with galena.

AMBLYGONITE
Possible Ore Mineral
Triclinic

of Lithium

$Li Al(FOH)PO_4$

Perfect one direction cleavage, fine parallel lines on face common; uneven fracture. Rough, poorly formed crystals, massive, compact cleavable masses. Usually white, though greenish and bluish known. Hardness 6 and brittle; specific gravity 3.0-3.1. Luster is vitreous, somewhat pearly on cleavage face, and is translucent. Yields water in closed tube. BB swells, fuses at 2 to white globule, produces red flame (Li), when powdered and moistened with sulphuric acid gives slightly bluish flame (P). Insoluble in acids, often fluorescent. Occurs in granite pegmatites; common associates lepidolite, spodumene, apatite, tourmaline.

DESCLOIZITE
Uncommon Secondary Mineral
Orthorhombic

of Lead

$Pb(Zn,Cu) VO_4(OH)$

Cleavage not clear, usually small crystals, drusy, also fibrous, sometimes mammillary surface. Platy, prismatic, radiating, fibrous. Dark brown to cherry red and black. Distinct orange to brownish-red streak. Adamantine luster. Hardness is 3.5 and specific gravity 5.9-6.2. Associates are galena, sphalerite, and copper carbonate and oxide minerals.

LAZULITE
Possible Gemstone
Monoclinic

$(Mg,Fe)Al_2(PO_4)_2(OH)_2$

Indistinct cleavage, uneven fracture, brittle. Sharp pyramids, also massive granular to compact. Sky blue, translucent to opaque, vitreous luster. Hardness is 5.0-5.5, and specific gravity 3.0-3.1. Infusible. Swells, decrepitates, loses color BB. Yields water in closed tube. Insoluble. Rare mineral, in pegmatites, quartz veins. Associates are rutile, kyanite, corundum, and sillimanite.

VANADINITE
(Coating)

AMBLYGONITE

DESCLOIZITE

5 cm

LAZULITE
(blue in Quartz)

WAVELLITE
Phosphatic Mineral — **Orthorhombic**

$Al_3(PO_4)_2(OH)_35H_2O$

Prismatic cleavage, uneven or fibrous fracture, brittle. Globular aggregates of radiating crystals. White, often greenish-yellow, gray, brown and green. Translucent and vitreous to silky luster. Hardness is 3.5-4.0 and specific gravity 2.3. Dissolves in HCl. Infusible, but will swell and split BB. Moistened with cobalt nitrate solution and heated intensely before BB turns blue (Al). Yields water in closed tube. On joint surfaces, in rock cavities, associated with phosphorite deposits.

TURQUOISE
Gemstone — **Triclinic**

$CuAl_6(PO_4)_4(OH)_85H_2O$
Faustite (Zn Isomorph)

No cleavage, uneven fracture and brittle. Rare minute crystals, mainly massive, granular, cryptocrystalline encrustations, veinlets. Sky blue to bluish-green to green (Faustite). Light green streak. Opaque and dull to waxy luster. Hardness 5-6 and specific gravity 2.6-2.8. Infusible. BB turns dark and decrepitates with vivid green flame (Cu). Yields water in closed tube. Soluble in HCl. Powdered mineral dissolves in HNO_3 acid. In warm ammonium molybdate yields yellow precipitate (PO_4).

AUTUNITE
Ore Mineral of Uranium — **Orthorhombic**

$Ca(UO_2)_2P_2O_88H_2O$

Good basal and prismatic cleavage, brittle laminae. Thin tabular, square crystals, micaceous. Lemon to sulphur yellow with pearly to subadamantine luster. Yellowish streak. Hardness 2.0-2.5 and specific gravity 3.1. Radioactive. Yields water in closed tube, fuses at 2.5 BB to blackish mass. Green bead in salt of phosphorous. Fluoresces. Soluble in nitric acid. Associated with torbernite its copper analog, and with other secondary uranium minerals.

TORBERNITE
Ore Mineral of Uranium — **Orthorhombic**

$Cu(UO_2)2P_2O_812H_2O$

Perfect basal cleavage, micaceous, also prismatic. Square, tabular, foliated. Laminae brittle. Emerald grass green, also apple and siskin green. Transparent to translucent, with pearly to subadamantine luster. Pale green streak. Hardness 2.0-2.5 and specific gravity of 3.2. Radioactive. Yields water in closed tube, fuses at 2.5 to blackish mass. Green bead in salt of phosphorous BB. Soluble in HNO_3. A secondary uranium mineral associated with autunite and other uranium minerals.

VIVIANITE
Phosphatic Mineral — **Monoclinic**

$Fe_3(PO_4)_2 8H_2O$

Micaceous cleavage with thin layers flexible. Tabular, prismatic, often radiating, reniform, fibrous, earthy. Fibrous fracture and also earthy, somewhat sectile. Colorless fresh, blue or green when oxidized. Transparent to translucent, when exposed to light becomes opaque. Has vitreous luster which is pearly on cleavage planes. White to blue-brown streak. Hardness is 1.5-2.0 and specific gravity 2.6-2.7 Fuses at 2 to black magnetic globule. Powder dissolves in HNO_3, add warm ammonium molybdate to solution and yellow precipitate occurs (PO_4). Yields water in closed tube. Rare, secondary, with copper veins, associated with pyrite, pyrrhotite, limonite.

WAVELLITE

TURQUOISE

VIVIANITE (blue)
on Quartzite (brown)

TORBERNITE

AUTUNITE

CARNOTITE   Ore Mineral of Uranium   **Orthorhombic**
$K_2(UO_2)_2(VO_4)_2 3H_2O$ and Vanadium

Basal perfect cleavage, but rarely seen, earthy fracture. Usually powdery, rarely thin tabular crystals. Bright yellow to greenish-yellow sometimes stained greenish or brownish. Earthy luster and opaque to translucent. Streak yellow. Hardness is 1.0-1.5, somewhat crumbly and specific gravity is 3.5-5.0. BB infusible. Dissolves in acids. Non-fluorescent, radioactive. Combined with hot bead of salt of phosphorus and reheated BB, bead turns yellowish-green in oxidizing flame and green in RF. Both this bead and borax bead fluoresce in UV light. Secondary mineral and resembles tyuyamunite, the calcium uranium vanadate.

ANNABERGITE   Secondary Nickel Mineral   **Monoclinic**
$Ni_3AsO_4 8H_2O$

Perfect cleavage. Apple green. Hardness is 2.5-3.0 and specific gravity 3.0. Fusible BB. A secondary mineral with smaltite and other nickel-bearing minerals. M.T. for Ni and As.

ERYTHRITE   Cobalt Mineral Indicator   **Monoclinic**
$CO_3(AsO_4)_2 8H_2O$ (Cobalt Bloom)

Perfect micaceous cleavage, and is sectile. Prismatic, acicular, radiating groups, uniform masses, powdery coatings. Crimson red to pink with vitreous to dull luster, on cleavage pearly. Streak is like mineral color but paler. Hardness is 1.5-2.5 and specific gravity 3.0-3.1. BB fuses at 2 and yields gray bead. Dissolves in HCl and produces red solution. In closed tube gives water. Rare mineral associated with cobalt minerals as cobaltite and skutterudite in weathering zone. M.T. 1-A.

TYUYAMUNITE   Ore Mineral of Uranium   **Orthorhombic**
$CaO-2UO_3-nH_2O$

Perfect prismatic cleavage, other distinct. Scales and earthy. Yellow to greenish-yellow, translucent. Relatively soft and specific gravity is 3.7-4.3. Dissolves in acids. Fluoresces yellow. Radioactive. Commonly associated with carnotite, its potassium analog. Of secondary origin from action of meteoric waters on primary uranium and vanadium minerals.

CARNOTITE

ANNABERGITE

TYUYAMUNITE (yellow)

ERYTHRITE (pinkish coating)

5 cm

WOLFRAMITE — Tungsten Ore Minerals — Monoclinic

$(FeMn)WO_4$

Includes series:

Ferberite - $FeWO_4$
Hubnerite - $MnWO_4$

Perfect one direction cleavage, uneven fracture, and brittle. Commonly tabular crystals, striated, columnar, lamellar, granular. Black to dark brown, opaque and resinous to submetallic. Dark brown to black streak. Hardness 5.0-5.5 and specific gravity 7.1-7.5. Fuses 2-4 BB to globule, Fe-bearing series is slightly magnetic. H_2SO_4 decomposes powdered mineral and colorless solution turns blue with addition of zinc. Uncommon, in veins with quartz, sulphide minerals, and in placers. M.T. 1-B.

SCHEELITE — Tungsten Ore Mineral — Tetragonal

$CaWO_4$

Powellite - $CaMoO_4$

Fairly distinct pyramidal cleavage, with uneven fracture and very brittle. Usually bipyramidal crystals, also massive and granular. White, sometimes shades of yellow, green, brown and red. Transparent to translucent with vitreous to adamantine luster. White streak. Hardness is 4.5-5.0 and specific gravity 5.9-6.1 (Powellite, H 3.5-5.0, G 4.2-6.1). Fluoresces blue to yellow-white depending on amount of molybdenum present. Fuses at 5. Decomposes in HCl. Powder boiled in HCl leaves yellow residue, and tungsten can be confirmed by adding tin, continuing boiling with solution turning blue and later brown. Rare mineral in contact metamorphic zone replacing carbonate rocks. Also in veins in association with quartz, wolframite, cassiterite, molybdenite, topaz, fluorite, and apatite.

WULFENITE — Possible Ore Mineral of Molybdenum — Tetragonal

$PbMoO_4$

Distinct pyramidal where visible, uneven to conchoidal fracture, brittle. Square plates or tablets, thin, also granular. Orange-yellow, olive green, brown. Translucent and resinous or adamantine luster. White streak. Hardness 3 and specific gravity is 6.8. Fuses at 2. Powder in sodium carbonate bead BB yields lead globule on charcoal. Powder in iodide flux (KI+S) heated on plaster of paris tablet yields deep blue coating (Mo). Rare, secondary mineral in oxidized zone of lead deposits. Associated with cerussite, pyromorphite, galena.

CROCOITE — Lead Chromate Mineral — Monoclinic

$PbCrO_4$

Distinct cleavage, with uneven to conchoidal fracture. Columnar, prismatic, granular. Sectile. Bright hyacinth red color, adamantine to vitreous luster, translucent. Hardness is 2.5-3.0 and specific gravity 5.9-6.1. Streak is orange-yellow. Fuses at 1.5, on charcoal block reduced to metallic lead in RF leaving a chromium oxide residue. Emerald green bead in salt of phosphorus. A secondary mineral associated with lead-bearing veins.

WOLFRAMITE (bladed crystals)
in Quartz

SCHEELITE

CROCOITE

WULFENITE

5 cm

138

TORBERNITE*   A Secondary Uranium Mineral   Orthorhombic
$Cu(UO_2)(PO_4)_2 8H_2O$ Pseudo-Tetragonal

See page 132 for description.

AUTUNITE*   A Secondary Uranium Mineral   Orthorhombic
$Ca(UO_2)_2(PO_4)_2 10 H_2O$ Pseudo-Tetragonal

See page 132 for description.

BECQUERELITE*   A Secondary Uranium Mineral   Orthorhombic
$(2UO_3 3H_2O)$

Perfect 001 cleavage, also 101. Tabular, prismatic, massive habit with striae on 100 and 010 faces. Amber to brownish yellow color, generally transparent, and adamantine to greasy luster. Yellow streak. Hardness is 2.0-3.0 and specific gravity 5.2. Radioactive. An oxidation product of uraninite, and associated with other brightly-colored secondary uranium minerals. Appropriatley named after A. Henri Becquerel, discoverer of radio-activity. Known from old Belgium Congo, Bavaria, in Great Bear Lake, Canada and in other uranium districts.

COLUMBITE*   Niobium and Tantalum   Orthorhombic
$(FeMn)(NbTa)_2O_6$ **Ore Minerals**
TANTALITE
$(FeMn)(TaNb)_2O_6$

Distinct cleavage, uneven fracture. Common crystals, heart-shaped twins, prismatic, thin tabular. Black, very shiny on fresh fractures. Generally has bluish tarnish. Submetallic to resinous luster and opaque to translucent. Streak dark brown to dark red and black. Hardness 6 and brittle, with specific gravity of 5.0-8.0. BB infusible, very slowly soluble in acids. Mixed with sodium carbonate and heated BB yields a magnetic mass (Fe). Uncommon mineral, occurs in pegmatites, with quartz, feldspar, tourmaline, beryl, spodumene, etc., and in placer deposits.

URANINITE*   Uranium Ore Mineral   Isometric
(Pitchblende) UO_2

No visible cleavage, uneven to conchoidal fracture, brittle. Rare crystals, massive, pitchblende more crystalline, botryoidal, lamellar and curved. Black, opaque, and pitch-like to dull submetallic luster. Brownish-black streak. Hardness is 5.5 and specific gravity 6.5-8.5 for pitchblende and 9.0-9.7 for uraninite. Very radioactive. Soluble in HNO_3 and H_2SO_4 acids. Infusible. Powdered mineral in hot bead of salt of phosphorous reheated BB produces yellowish green (oxidizing flame) and green (reducing flame). Both this bead and borax bead fluoresce in UV light, the mineral will not. Occurs in granite pegmatites, in veins, sandstone environments with associates of lead, silver, copper, nickel, zinc, and other elements.

GUMMITE*   Secondary Uranium Mineral   Not Clearly Defined
(UO_2NH_2O) **Possible mixtures of Curite (a lead uranium oxide)**
                          and alkaline- and RE, Fe, Al, and P-oxides

Cleavage not apparent, and uneven to conchoidal fracture. Quite brittle, yellow, orange, reddish-yellow to orange-red, brown to black. Yellow streak. Greasy, waxy, vitreous to gum-like luster. Small grains transparent. Hardness is 2.5-5.0. Specific gravity is 3.9-6.4 depending on composition. Occurs at many uraninite localities in U.S. in uranium-bearing pegmatites in North Carolina, New Hampshire, Connecticut, and Maine; also at Great Bear Lake district, NWT. Found as pseudomorph after uraninite, in pegmatites possibly as result of late-stage hydrothermal solutions.

* Courtesy of Utah Museum of Natural History.

AUTUNITE

TORBERNITE on MALACHITE

COLUMBITE-
TANTALITE

BECQUERELITE

5 cm

URANINITE (black)
with GUMMITE

WILLEMITE **Rare Ore Mineral of Zinc** **Hexagonal**
Zn_2SiO_4

Good basal cleavage, uneven to conchoidal fracture, brittle. Prismatic, usually massive, fibrous or grains. Greenish-yellow typically but varies from white to brown. Transparent to opaque and vitreous to resinous luster. Hardness is 5.5 and specific gravity 3.9-4.2. Infusible. Soluble in HCl. Fluoresces green. Powder in sodium carbonate bead BB on charcoal yields yellow coating (hot) and white (cold). Moisten coating with cobalt nitrate solution and heat BB and it turns green (Zn). Rare, in zinc deposits with zincite, franklinite, calcite.

OLIVINE **Rock-forming Mineral** **Orthorhombic**
$(Mg,Fe) SiO_4$
Forsterite $(MgSiO_4)$
Fayalite $(FeSiO_4)$

Indistinct cleavage, conchoidal fracture and brittle. Isolated grains in igneous rock, aggregates, crystals rare. Green, olive-green, sometimes brownish, grayish-red. Transparent to translucent and vitreous luster. Hardness is 6.5-7.0 and specific gravity 3.3-3.4. Infusible. Iron variety (fayalite) magnetic after heating BB in reducing flame and cooling. Powder may dissolve in concentrated HCl. Continue boiling and silica gel forms when test tube is nearly dry. Associated with augite, chromite, magnetite, serpentine, garnet and spinel.

PHENACITE **Possible Gemstone** **R-Hexagonal**
Be_2SiO_4

Poor prismatic cleavage, conchoidal fracture, brittle. Small rhombohedral crystals, some lens-shaped and prismatic. Colorless to white sometimes yellowish, pink, brown. Transparent to translucent. Vitreous luster. Hardness is 7.5-8.0 and specific gravity 3.0. Infusible and insoluble. Rare mineral in openings in granite and in pegmatites. Associates are topaz, beryl, apatite, chrysoberyl.

ZIRCON **Possible Gemstone** **Tetragonal**
$ZrSiO_4$

Indistinct cleavage, conchoidal fracture and brittle. Square prisms or pyramids, also rounded grains. Generally brown, also colorless, grayish, orange, yellow-green, red, blue rarely. Translucent to transparent, often opaque with adamantine luster. Hardness is 7.5 and specific gravity 4.7. Insoluble and infusible. BB may glow or turn white. Powdered mineral fused with sodium carbonate and dissolved in dilute HCl turns tumeric paper orange in solution. May fluoresce orange. Common, widespread accessory mineral. In igneous rock, mainly granite, syenites, and diorite.

PHENACITE

WILLEMITE (dark)

OLIVINE

ZIRCON (dark crystals)

5 cm

ZIRCON

GARNET (Group)		**Possible Gemstone if**	**Isometric**
$Fe_3Al_2(SiO_4)_3$	Almandite	**Transparent and**	
$Ca_3Fe_2(SiO_4)_3$	Andradite	**Good Color**	
$Ca_3Al_3(SiO_4)_3$	Grossularite		
$Mg_3Al_2(SiO_4)_3$	Melanite (like andradrite)		
	Pyrope		
	Rhodolite (2 parts pyrope)		
$Mn_3Al_2(SiO_4)_3$	Spessartite		
$Ca_3Cr_2(SiO_4)_3$	Uvarovite		

Indistinct cleavage and uneven fracture. Dodecohedrons, trapezo-hedrons or combination, also massive. Vitreous to resinous luster and trans-parent to translucent. Hardness is 6.5-7.5 and specific gravity 3.4-4.3. BB fuss at 3-4 except infusible uvarovite. Insoluble in acids. Common mineral in metamorphic rocks including marble, and in igneous rocks, pegmatites and in sands as a heavy mineral.

The garnet group consists of several varieties based on chemical composition. All crystallize in the isometric system, and have similar habits of crystallization mainly as dodecahedrons and trapezohedrons. The same general chemical composition applies though the elements present do differ.

The aluminum-bearing garnets include: grossularite, pyrope, alman-dite and spessartite. The iron garnet group consists of andradite, the calcium iron silicate, but may have magnesian, titanium, or yttrium as substitutions. A third variety includes the chromium-bearing garnet uvarovite.

Several variety names are used for the garnet group. These are often based on color, which in turn is a reflection of the chemical composition. Hessonite, or cinnamon-stone, known as hyacinth from Sri Lanka, resembles, but is harder than the yellow and yellowish red types. Normal grossularite is colorless to white, pale green, amber and honey yellow. Specific gravity is 3.5.

Pyrope, the magnesium aluminum garnet, is deep red to nearly black. If perfectly transparent, it is used as a gem. Rhodolite is a close relative of pyrope and has delicate shades of rose red and purple. Specific gravity is 3.5.

Almandite may be a precious mineral of the iron-aluminum variety. Colors are fine deep red, brownish red, black. The red transparent types are gem varieties. Specific gravity is 4.3.

Spessartite is the manganese-aluminum garnet of 4.2 specific gravity and is dark hyacinth red, sometimes with tints of violet and brownish red.

Andradite is a common garnet of calcium, iron silicate composition with substitutions of the ferrous iron by aluminum, while ferrous iron, man-ganese and sometimes magnesium replace the calcium. Colors likewise vary from topaz and greenish yellow, apple green to emerald green, reddish, grayish green, dark green and brown to black. Specific gravity is 3.7+. Topazolite is a variety with a topaz color and transparency. The grass green to emerald green type is demantoid and has a brilliant luster for gem use. The resinous luster variety is colophonite, named from the resin colo-phony. Melanite is black and is rather dull and lustrous.

The manganesian calcium-iron garnet includes several varieties. Most important is allochroite, a manganesian iron-garnet of brown to reddish brown color. The titaniferous garnet approaches schloromite and is black like the tourmaline schlorite, while the yttrium-bearing garnet is rare.

Uvarovite is a calcium-chromium garnet of 7.5 hardness and specific gravity of 3.4-3.5. It is a distinctive emerald green.

GARNET GROUP

PYROPE

GROSSULARITE

MELANITE

ALMANDITE

ALMANDITE

SPESSARTITE

5 cm

ANDRADITE

SILLIMANITE   Some Gem Quality   Orthorhombic
Al_2SiO_5

Good pinacoidal cleavage, uneven fracture and brittle. Elongated prismatic, fibrous, interwoven, radiating. Colorless, white, grayish, yellowish, brownish, sometimes greenish. Transparent to translucent and vitreous luster. Hardness is 6-7 and specific gravity 3.2. Insoluble and infusible. Powder moistened with cobalt nitrate solution gives blue of aluminum when heated intensely BB. Uncommon, generally in schists and gneisses associated with corundum, andalusite, cordierite.

ANDALUSITE   Clear, Transparent,   Orthorhombic
Al_2SiO_5 Possible Gemstone

Prismatic cleavage, but not prominent; fracture is uneven. Prismatic, pseudo-tetragonal, square x-section, and massive. Pink, reddish brown, olive green, pale grey, and white. Color varies on direction of view, generally green or red. Vitreous luster, often dull, may be transparent to opaque. Hardness 7.5 and is brittle; specific gravity 3.2. Infusible and insoluble. BB powdered mineral moistened with cobalt nitrate solution and heated intensely turns blue (Al). A common mineral in metamorphic rocks, especially contact metamorphic zones. Associated minerals are kyanite, sillimanite, garnet, corundum, and tourmaline.

KYANITE   Ceramic Use Mineral   Triclinic
Al_2SiO_5

Two good cleavages and brittle. Long bladed crystals, also radiating blades. Blue, also green, black, white or colorless, sometimes spotty. Darkest at center of crystal. Vitreous to pearly luster, transpaent to transiucent. Hardness 5 parallel to length, 7 across crystal. Specific gravity 3.6-3.7. Infusible BB and insoluble. Moistened with cobalt nitrate solution and heated BB turns dark blue. Associated minerals corundum, staurolite, and garnet in dynamic metamorphosed rocks.

STAUROLITE   Possible Gemstone   Pseudo-orthorhombic
$FeAl_4Si_2O_{10}(OH)_2$ If Transparent and Clear

One distinct cleavage, subconchoidal fracture, brittle. Prismatic, rarely massive, commonly cross-line twinning. Brown to brownish black. Transparent and opaque and vitreous to resinous luster. Hardness 7.0-7.5 and specific gravity 3.7-3.8. Insoluble and infusible. Rare in schists, gneisses and slates. Garnet, kyanite, tourmaline are associates.

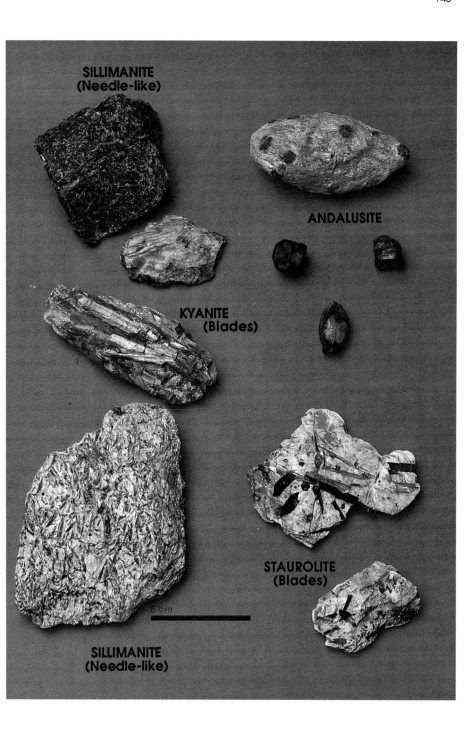

SILLIMANITE
(Needle-like)

ANDALUSITE

KYANITE
(Blades)

STAUROLITE
(Blades)

5 cm

SILLIMANITE
(Needle-like)

TOPAZ   Gemstone   Orthorhombic
$Al_2SiO_4(F,OH)$

Perfect one direction cleavage, uneven fracture and brittle. Pointed prisms, striated columnar, and granular. Mainly colorless, also white, pale yellow, brown, reddish, green and bluish. Transpaent to translucent and good vitreous luster. Hardness 8 and specific gravity is 3.5-3.6. Insoluble and infusible. Powdered mineral moistened with cobalt nitrate solution and heated BB turns blue (Al). Rare, in pegmatites, cavities in silicic volcanic flows, and metamorphic rocks, and in placers. Associates are cassiterite, fluorite, tourmaline, and quartz.

CHONDRODITE   Monoclinic
$2Mg_2SiO_4Mg(F,OH)_2$

Sometimes distinct cleavage, with uneven to subconchoidal fracture and brittle. White, light yellow, honey yellow dark brown and red. Vitreous to resinous luster. Translucent. Hardness is 6-7 and specific gravity 3.1-3.2. Infusible BB; may blacken and burn white. Potassium bisulphate fusion in closed tube shows reaction for fluorine. May gelatizine with acids. Occurs in metamorphosed dolomitic limestones near igneous bodies. Associates are phlogopite, spinel, pyroxene, and magnetite.

BERTRANDITE   An Ore Mineral of   Orthorhombic
$H_2Be_4Si_2O_7$ Beryllium

Perfect prismatic cleavage and brittle. Colorless to pale yellow, translucent. Vitreous luster. Hardness is 6-7 and specific gravity 2.6. A rare mineral mainly in pegmatites but found in quantity as small crystals, and replacements in altered volcanic tuff near Topaz Mountain, Utah, the world's premier source of beryllium.

DATOLITE   A Boron-bearing Mineral   Monoclinic
$HCaBSiO_5$

Cleavage not observed, conchoidal fracture and brittle. White, sometimes grayish, pale green, yellow, red, purplish, rarely dirty green or honey yellow. Vitreous luster, less so for subresinous or fracture surface. Has a white streak. Specific gravity is 2.9-3.0 and hardness 5.0-5.5. Gives off water in closed tube. BB fuses at 2 with swelling to clear glass, yields bright green flame. In HCl gelatinizes. A secondary mineral in veins and cavities in basic igneous rocks. Mineral associates are calcite, prehnite, and zeolites; also in gneiss, diorite, serpentine, and metallic-bearing veins.

TOPAZ

CHONDRODITE (brown)
in Dolomite

DATOLITE

BERTRANDITE in
FLUORITE

5 cm

SPHENE
CaTiSiO₅

**Possible Ore Mineral of Titanium
Common Accessory Mineral in
Igneous Rocks**

Monoclinic

Good prismatic cleavage, conchoidal fracture and brittle. Flattened and wedge-shaped crystals, also massive. Commonly brownish and greenish yellow, sometimes gray and black. Transparent to opaque and resinous to adamantine luster. Hardness is 5.0-5.5 and specific gravity 3.4-3.6. Fuses 4 to dark mass. On charcoal fuse powder with sodium carbonate, dissolve, mix in concentrated HCl, add tin and boil until light violet color appears (Ti). H_2SO_4 decomposes mineral, only slightly decomposed by HCl. Common, accessory mineral in igneous rocks.

DUMORTIERITE
$Al_8BSi_3O_{19}(OH)$

Silicate Mineral

Orthorhombic

Poor cleavage and uneven to conchoidal fracture. Quite brittle. Rare crystals, usually fibrous, radiating. Blue, violet, pink with vitreous luster, and transparent to translucent. Hardness of 7.0, distinct, and specific gravity of 3.3. BB infusible but turns white. Powder BB on charcoal and moistened with cobalt nitrate gives blue (aluminum). In gneisses, schists and some pegmatites.

HEMIMORPHITE
$Zn_4Si_2O_7(OH)_2H_2O$

Zinc Ore Mineral

Orthorhombic

Prismatic cleavage perfect when observable. Uneven fracture. Tabular, radiating, fibrous, some mammillary. White, pale blue, pale green, yellowish brown. Transparent to translucent, vitreous luster. Hardness 4.5-5.0 and specific gravity 3.4-3.5. BB fuses at 5, decrepitates, whitens, yields water in closed tube. Mix powder with sodium carbonate and heat BB, yields yellow coating hot and white when cold. Moisten coating with cobalt nitrate and mineral with same, reheat BB and coating turns green and mineral blue (Zn). Oxidized mineral of zinc, associated with smithsonite ($ZnCO_3$) and sphalerite (ZnS).

EPIDOTE
$Ca_2(Al,Fe)_3(SiO_4)_3OH$

Common Silicate

Monoclinic

One direction perfect cleavage, uneven fracture, brittle. Prismatic, striated, also granular, and fibrous. Pistachio green, blackish green. Vitreous luster and transparent to opaque. Streak grayish to colorless. Hardness is 6-7 and specific gravity 3.3-3.5. BB fuses at 3-4, forms swelled black mass which is commonly magnetic. Powder yields water in closed tube when intensely heated. Common associates are actinolite, feldspar, hornblende, quartz, and chlorite.

SPHENE

DUMORTIERITE

HEMIMORPHITE

EPIDOTE

5 cm

ALLANITE Rare-Earth bearing Mineral **Monoclinic**
Epidote-like with Ce,
Ih, La   lesser Y

Two-direction cleavage, uneven fracture to conchoidal, and brittle. Brown
to black, pitchy, resinous to submetallic luster, and subtranslucent to
opaque. Hardness is 5.0-5.5 and specific gravity 3.0-4.2. Yields water in
closed tube. Fuses easily BB with swelling to dark, blobby, magnetic glass.
May gelatinize in HCl pre-ignition. An accessory mineral in plutonic rocks.
Also in orthogneisses, amphibolite, and contact metamorphosed limestone.

IDOCRASE Silicate Mineral in Contact **Tetragonal**
(Vesuvianite) Metamorphic Setting
$Ca_{10}Al_4(Mg,Fe)_2Si_9O_{34}(OH)_4$

Poor to distinct cleavage, subconchoidal to uneven fracture. Usually short
square prisms, striated, columnar, massive, granular. Dark green, brown to
yellowish, sometimes blue (cyprine). Vitreous to resinous luster, subtransparent to translucent. Hardness is 6.5 and specific gravity 3.3-3.4. BB fuses at 3.
Swells and fuses to brown or green non-magnetic glass. Yields water in
closed tube. HCl slightly decomposes it. Uncommon mineral in contact
metamorphosed limestone. Associated with garnet, wollastonite, epidote,
diopside. Also called vesuvianite, named from Mt. Vesuvius location.

BERYL Ore Mineral of Beryllium And **Hexagonal**
$Be_3Al_2Si_6O_{18}$ Gem Minerals: Emerald, Aquamarine,
Morganite, Golden Beryl (Heliodor),
Goshenite, Rösterite and Vorobyevite

Poor cleavage and conchoidal to uneven fracture. Pale green, yellow, pink,
bluish white and colorless. Emerald green-transparent is _emerald_, and
greenish blue-transparent is _aquamarine,_ while pink to rose transparent is
morganite, and clear yellow is _golden beryl_, white is goshenite, and white
Na/Li beryl is rösterite, while high alkali beryl of white, pink, pale blue is
vorobyevite. Vitreous luster, transparent to translucent. Hardness is 7.5-8.0,
and specific gravity 2.8. BB fuses on thin edges or splinters, turns white.
Insoluble. Commonly in pegmatites, also mica schists, placer deposits.
Quartz, feldspar, garnet, tourmaline, mica, spodumene mineral associates.
Important localities in Maine, Pennsylvania, and Connecticut (golden);
Bogota, Columbia, Alexander Co., N. C. (emerald).

CORDIERITE Sometimes a Gemstone **Orthorhombic**
$(Mg,Fe)_2Al_4Si_5O_{18}$

Distinct prismatic and indistinct basal cleavage with subconchoidal fracture and brittle. Shades of blue, light, dark and smoky blue. Vitreous luster
and transparent to translucent. Hardness 7.0-7.5 and specific gravity is 2.6-
2.7. BB fuses at 5.0-5.5 loses transparency. Partly decomposed by acids.
Found in gneisses, schists, and contact metamorphic zones, also in granites,
and silicic to intermediate volcanic rocks.

ALLANITE

IDOCRASE

MASSIVE BERYL

CORDIERITE

5 cm

BERYL

TOURMALINE   **Silicate**   **Hexagonal**

$Na(Mg\ Fe\ Li\ Al\ Mn)_3Al_6(BO_3)_3Si_6O_{18}(OH,F)_4$

Several varieties:

    Schorl-black   Indicolite-blue
    Dravite-brown   Brazilian Emerald-green
                     Rubellite-pink

No visible cleavage, uneven to conchoidal fracture, brittle. Slender, vertically striated prisms with triangular cross section, also massive. Black common, brown, also green, blue, red. May have zoned colors. Transparent to opaque and vitreous luster. Hardness 7.0-7.5 and specific gravity 3.0-3.2. Fuses 3 to infusible. Acid insoluble. Some fluoresces. Common mineral; usually in pegmatites, igneous rocks, and metamorphic rocks. Associates include quartz, feldspars, micas, topaz, beryl, apatite, and fluorite. Excellent crystals at Paris, and Auburn, Maine., also Pala, California, in Brazil's Ural Mountains., Russia, and Madagascar.

CHRYSOCOLLA   **Copper Ore Mineral**   **Cryptocrystalline**

$CuSiO_3$

No cleavage visible, has conchoidal fracture. Often botryoidal, enamel-like crusts and seams. Blue to greenish, but impure variety may be brownish or blackish. Vitreous to earthy luster and translucent to opaque. White streak. Hardness is 2-4, somewhat brittle to sectile. Specific gravity is 2.0-2.4. Sticks to tongue. BB infusible, yields water in closed tube. Mix with sodium carbonate and BB gives green (Cu) flame. Decomposed by acids. Occurs in oxidized portions of copper deposits associated with azurite, malachite, cuprite, chalcopyrite. M.T. 1-A.

ENSTATITE   **Rock-forming Mineral**   **Orthorhombic**

$MgSiO_3$ (Bronzite)

Good prismatic cleavage and basal parting, uneven fracture and brittle. Grayish, yellowish or greenish white to olive green an brown. Uncolored streak. Translucent to almost opaque with vitreous to pearly on cleavage surface. Somewhat metallic luster in bronzite. Hardness is 5.5 and specific gravity 3.1-3.3. Almost infusible BB (6.0). Insoluble in HCl. Common constituent of igneous rocks low in lime, sometimes in metamorphic rocks. Best observed in diorites, gabbros, norites and pyroxenites and peridotites and surface equivalents.

DIOPSIDE   **Member of**   **Monoclinic**

(Ca, Mg, Fe, Tl, Al)   **Diopside-Hedenbergite**
                       Series Clinopyroxene

Good prismatic cleavage, sometimes basal parting visible. Stout prisms, 4- or 8-sided in section, also massive, granular. Light to dark green to black, translucent to opaque with vitreous luster. Specific gravity is 3.2-3.6 and hardness 5.5-6.5. Common mineral in darker colored igneous rocks, also in metamorphic rocks, contact metamorphic rocks in skarn.

TOURMALINE (Schorl)

CHRYSOCOLLA

TOURMALINE
(Schorl)

DIOPSIDE

ENSTATITE

5 cm

SPODUMENE
LiAlSi$_2$O$_6$

Gemstones
Hiddenite and Kunzite Varieties

Monocline

Perfect prismatic cleavage with uneven fracture to subconchoidal and brittle. Greenish white, grayish white, yellowish green and emerald green (Hiddenite), yellow, red, lilac (Kunzite, transparent), amethystine purple. Transparent to translucent with vitreous luster, pearly on cleavage planes. Streak white. Hardness is 6.5-7 and specific gravity 3.1-3.2. Swell BB; and becomes white and opaque. Yields purple-red flame, fuses 3.25 to clear or white glass. Non-reactive to acids. Occurs in granite pegmatites, sometimes in very large crystals to feet in length.

RHODONITE
MnSiO$_3$

Ore Mineral of Manganese

Triclinic

Perfect prismatic cleavage at 88° and 92°, a third less distinct cleavage. Prismatic, tabular, also massive , compact, fibrous. Uneven fracture and brittle. Pink, rose-red. May stain brown or black. Transparent to translucent with vitreous luster. Hardness is 5.5-6.5 and specific gravity is 3.4-3.7. Fuses 3 to dark glass. Powder in borax bead heated BB in oxidizing flame yields violet (hot) and reddish violet (cold). Uncommon mineral, associated with manganese minerals, calcite.

JADEITE
NaAlSi$_2$O$_6$

Possible Gem Mineral

Monoclinic

Poor prismatic cleavage at right angles, splintery fracture. Greenish to dark greenish, mottling of white and green. Vitreous luster, pearly on cleavage surfaces. Translucent to opaque. Very difficult to break. Hardness 6.5-7, and specific gravity 3.3 - 3.5. Rare crystals in compact, fibrous masses. BB fuses easily at 2.5 and forms transparent, bubbly glass. Yellow flame, characteristic of sodium. Insoluble. Rare mineral, may be in boulders in gravels from weathered and eroded serpentine areas.

AUGITE
Ca(MgFeAl)(AlSi)$_2$O$_6$

Rock-forming Silicates
Augite-black,
Diopside-pale green

Monoclinic

Good prismatic cleavage, may have basal parting, uneven fracture and brittle. White to green, pale green is diopside. Dark green to black – augite. Both have vitreous luster, are transparent to translucent, latter true for augite on thin edges. Streak to gray. Hardness is 5-6 and specific gravity 3.2-3.5. Fuses (4) to glass, diopside nearly colorless and augite black insoluble. Usually stout prisms, 4- or 8-sided cross section, also granular, massive. Common minerals, diopside in contact metamorphic environment with epidote, idocrase, garnet, tremolite. Very common rock-forming mineral in igneous rocks of low quartz content.

WOLLASTONITE
CaSiO$_3$

Silicate

Monoclinic

Two perfect cleavages at 84 and 96 degrees, uneven to splintery fracture and brittle. Colorless, white, grayish, reddish. Translucent and silky to vitreous luster. Hardness 4.5-5.0 and specific gravity 2.8-2.9. Decomposes in HCl, slight effervescence not uncommon. Common contact mineral in limestone – igneous rock contact zone. Tabular, prismatic, massive to fibrous. Associates are tremolite, epidote, garnet, diopside, idocrase.

SPODUMENE

RHODONITE with
MANGANESE OXIDE

(KUNZITE)

JADEITE

WOLLASTONITE

AUGITE

5 cm

PREHNITE   Silicate   Orthorhombic
$Ca_2Al_2Si_3O_{10}(OH)_2$

Good basal cleavage, uneven fracture, brittle. Curved tabular, reniform, rough crystalline, fibrous. Generally light green, white, sometimes yellow. Vitreous luster, transparent to translucent. Hardness is 6.0-6.5 and specific gravity 2.8-3.0. Fuses and swells at 2.5 to bubbly glass. Yields water in closed tube. Common secondary mineral in cavities in basalts and dark igneous rocks. Associates include datolite, pectolite, calcite, zeolites, native copper.

HYPERSTHENE   Rock-forming Mineral   Orthorhombic
$(Fe, Mg)SiO_3$

Distinct prismatic, some parting with uneven fracture and brittle. Massive, fibrous, lamellar rare, prismatic or tabular crystals. Dark brownish green, grayish black, greenish black and brown. Grayish to brownish gray streak. Translucent to opaque and metalloidal to somewhat pearly luster. Hardness is 5-6 and specific gravity 3.4-3.5. Fuses to black enamel BB, more easily with higher iron content. Yields magnetic mass on charcoal BB. HCl acid partially decomposes mineral. Occurs in basic igneous rocks as norite, gabbro, and more often in extrusive rocks such as andesite and trachyte.

PECTOLITE   Silicate Mineral   Triclinic
$Ca_2NaSi_3O_8(OH)$

Perfect cleavage, yields needle-like masses, splintery fracture and brittle. Radiating masses, sharp, needle-like crystals. Whitish or grayish, subtranslucent to opaque, silky luster. Hardness is 5 and specific gravity 2.7-2.8. Fuses at 2.3 to white glass, produces yellow (Na) flame. Is decomposed by HCl, may fluoresce. Uncommon mineral in fractures, cavities in basic igneous rocks. Common associates are prehnite, natrolite, datolite, apophyllite, and calcite.

TREMOLITE   Silicate   Monoclinic
$Ca_2Mg_5(Si_8O_{22})(OH)_2$
Nephrite (Jade)

Prismatic cleavage, bladed, massive, and fibrous. White to gray and transparent to translucent. Vitreous luster. Hardness 5-6 and specific gravity of 2.9-3.2. Non-reactive to HCl. Grades into actinolite with increase of iron content in place of magnesium. Compact. Fine-grained variety known as nephrite (jade). Generally characteristic of contact metamorphosed siliceous dolomites, also schists.

PREHNITE

HYPERSTHENE

PECTOLITE

TREMOLITE

5 cm

HORNBLENDE   A Common Silicate Mineral   Monoclinic
$Ca_2Na(Mg,Fe)_4(AlFeTi)_3Si_6O_{22}(O,OH)_2$

Perfect to prismatic, housetop-like cleavage, uneven fracture. Prismatic, diamond-shaped cross section, also bladed, columnar, fibrous, some massive. Light to dark green to black, vitreous luster, translucent on thin edges. Streak is white to pale green. Hardness is 5-6, quite brittle and specific gravity is 3.0 3.3. BB fuses at 4 to black globule, yields water in closed tube. Yellow flame (Na). Abundant mineral in igneous rocks, schists and gneisses. Common associates calcite, plagioclase, quartz, garnet.

ACTINOLITE   Rock-forming Mineral   Monoclinic
$Ca_2(Mg,Fe)_5(Si_8O_{22})(OH)_2$

Good prismatic cleavage, fracture uneven to subconchoidal, brittle. Bright green, grayish green. Sometimes transparent. Vitreous to pearly luster. Hardness is 5-6 and specific gravity 3.0-3.2. Fuses BB differently with iron content variable, about 2.5-4.0. Widely distributed rock-forming mineral in both igneous and metamorphic rocks. Actinolite is common in schist.

CUMMINGTONITE   Rock-forming Mineral   Monoclinic
$(Mg,Fe)SiO_3$ to Grunerite $(Fe,Mg)SiO_3$

Prismatic cleavage, housetop type; uneven fracture and brittle. Brown, silky luster. Hardness is 5-6, specific gravity 3.1-3.6 (grunerite).

GLAUCOPHANE   Rock-forming Mineral   Monoclinic
$Na(Al,Fe)(SiO_3)_2$

Perfect prismatic cleavage, uneven to conchoidal fracture and brittle. Azure blue, lavender blue, bluish black, grayish. Grayish blue streak. Translucent and with vitreous to pearly luster. Hardness is 6.0-6.5 and specific gravity 3.0-3.2. Occurs only in metamorphic rocks as in schists, crystalline marble and is associated with epidote, quartz, pyroxene, garnet, chlorite, and others.

ACTINOLITE

HORNBLENDE

GLAUCOPHANE

CUMMINGTONITE

5 cm

APOPHYLLITE Tetragonal
$KCa_4(Si_2O_5)_4 8H_2O$

Perfect one direction cleavage, fracture uneven. Varied habit, prisms, bipyramidal, and pinacoidal. Colorless, white, grayish and less commonly pale green or pink. Vitreous luster, pearly on base and transparent to translucent. Hardness 4.5-5.0, brittle, with specific gravity of 2.3-2.4. BB fuses at 2, swelling to white porous enamel. Yields water in closed tube, and is decomposed by HCl. Violet flame (K) when seen through blue glass. A fairly common mineral. Lines openings in basalt and other dark igneous rocks. Generally of secondary origin and is associated with zeolites, datolite, and pectolite.

KAOLINITE In Quantity and Pure State Monoclinic
$Al_2Si_2O_5(OH)_4$ Ceramic Use Mineral

Rarely seen, but perfect one direction cleavage, uneven fracture. Microscopic hexagonal plates, usually earthy masses. White, commonly stained brown, gray, and other colors. Earthy luster, opaque. Hardness is 1.0-2.5. Specific gravity 2.6 (difficult to assess). Insoluble, infusible. Earthy odor. BB turns blue (aluminum) when moistened with cobalt nitrate. Yields water in closed tube. Soft soapy feel between fingers. Common secondary mineral from aluminum silicates, especially feldspar.

SERPENTINE AND CHRYSOTILE Monoclinic
$Mg_3Si_2O_5(OH)_4$

No cleavage, conchoidal and splintery fracture and brittle. Crystals unknown, massive, platy and sometimes fibrous. Shades of green often mottled. Dull and waxy or greasy in massive form, silky when fibrous. Greasy feel. Translucent to opaque. Hardness is 2.5-5.0 and specific gravity 2.2-2.6. Infusible. Decrepitates BB and yields water in closed tube. Decomposed by HCl. Common secondary mineral in igneous and metamorphic rocks. Associates are magnesite, chromite, spinel, garnet, pyroxene, amphibole, and olivine.

GARNIERITE Nickel Ore Mineral
$(Ni,Mg)SiO_3 nH_2O$

Amorphous and friable and soft. Bright green, pale green to nearly white. Dull luster. Specific gravity is 2.3-2.8. May adhere to tongue. A secondary mineral in oxidized zone. Occurs in serpentine, also with other weathered ultra-mafic rocks. M.T. 1-A.

APOPHYLLITE

KAOLINITE

SERPENTINE

GARNIERITE

ASBESTOS
(Chrysotile)

5 cm

TALC Silicate Monoclinic
$Mg_3Si_4O_{10}(OH)_2$

Micaceous cleavage, flexible but not elastic layers, uneven or hackly fractures, sectile. Rare crystal, mostly foliated or massive. White, gray, green, brown. Transparent to opaque with pearly to greasy luster. Hardness 1-1.5 and specific gravity 2.7-2.8. Soapy feel. Fuses at 5 and whitens. Insoluble in acids. Powder moistened with cobalt nitrate solution and heated BB turns violet. Secondary mineral in metamorphic rocks mainly, also in igneous rocks.

PYROPHYLLITE Orthorhombic
$H_2Al_2(SiO_3)_4$

Good basal cleavage, laminae flexible, but not elastic. Foliated, radiating, fibrous, granular, compact. Greasy feel. White, apple green, grayish, yellow tones for color. Subtransparent to opaque. Hardness 1-2, specific gravity 2.8-2.9. BB whitens and fuses with difficulty. Moisten with cobalt solution and heat to show deep blue color indicative of aluminum, distinction from talc which it resembles. Occurs in general stress metamorphic environment.

URANOPHANE Ore Mineral of Orthorhombic
$CaO\text{-}2UO_3\text{-}2SiO_2\text{-}7H_2O$ Uranium

Cleavage in crystalline varieties. Radiating, fibrous, prismatic, massive. Yellow, transparent. Hardness is 2-3 and specific gravity 3.8-3.9. Radioactive. Commonly found as secondary mineral in granite. Associated minerals – secondary uranium minerals, also with fluorite and uraninite.

MUSCOVITE White Mica Monoclinic
$KAl_3Si_3O_{10}(OH)_2$ Rock-forming Mineral
Fuchsite (Green)

Basal perfect cleavage, plates flexible and elastic. Tabular crystals, scaly, foliated masses. Colorless, light tones of yellow, brown, green. Bright green, chrome bearing variety is known as fuchsite. Transparent to translucent, vitreous to pearly luster. Hardness is 2.5-3.0 and specific gravity 2.7-3.0. Fuses at 5, yields water in closed tube. Insoluble. Common rock-forming mineral in granite pegmatite and schist.

TALC

PYROPHYLLITE

MUSCOVITE
and
QUARTZ

(FUCHSITE)

MUSCOVITE

URANOPHANE in KAOLINITE

5 cm

BIOTITE Rock-forming Mineral Monoclinic
$K(Mg,Fe)_3AlSi_3O_{10}(OH)_2$

Perfect micaceous (elastic) cleavage, uneven fracture. Scales, foliated, tabular, short prismatic, and hexagonal. Black, brownish black, greenish black, vitreous luster, translucent to opaque. Hardness is 2.5-3.0. Specific gravity 2.7-3.1. BB fuses at 5. Yields water in closed tube. In boiling H_2SO_4 acid turns milky and mineral decomposes. Common mineral in igneous and metamorphic rocks. Light colored variety of mica is muscovite.

LEPIDOLITE Lithium-bearing Mica Monoclinc
$KLiAl_2Si_3O_{10}(OH,F)_2$

Perfect basal cleavage. Flexible and elastic. Scaly, granular, rarely prismatic crystals. Granular or scaly fracture. Lilac, colorless, grey or pale pink. Pearly luster and translucent. Hardness is 2.5-4.0 and specific gravity 2.8-3.3. BB fuses at 2, produces red (lithium) flame. In closed tube yields acid water. Insoluble in acids. Uncommon mineral, generally in pegmatites associated with spodumene, pink tourmaline, muscovite and amblygonite.

PHLOGOPITE Member of Mica Group Monoclinic
$H_2KMa_3Al(SiO_4)_3$

Perfect basal cleavage, laminae tough and elastic. Scales, plates, prismatic. Yellowish brown to brownish red, coppery reflection. Pearly luster, bright submetallic on cleavage surface. Translucent. Hardness 2.5-3.0 and specific gravity 2.8. Star-like image produced if candle is held behind plate. Much like biotite in other characteristics. Common in metamorphic limestone and dolomite, also in serpentine. Mineral associates are pyroxene and amphiboles.

CHLORITE Group Name for Clinochlore, Monoclinic
$(MgFe)_5Al_2Si_3O_{10}(OH)_8$ Prochlorite, Penninite, and Kammererite

Micaceous, one-direction perfect cleavage and uneven fracture. Tabular, pseudo-hexagonal, rarely prismatic, scaly, earthy. Green, also yellow, brown, red. Vitreous to pearly luster and translucent to opaque. Streak light green to colorless. Hardness is 2.0-2.5 and specific gravity 2.6-2.9. Flexible but not elastic thin sheets. BB fuses, difficultly, at 5.0-5.5 and yields water in closed tube under intense heating. A common, widespread mineral group. Occurs in schists, serpentine, other low-grade metamorphic rocks and igneous rocks.

LEPIDOLITE

BIOTITE

CHLORITE
(Kammererite)

PHLOGOPITE

5 cm

CHLORITE
(Clinochlore)

Quartz
SiO_2

Common Rock-forming Mineral
Select Varieties are Gemstones

Hexagonal

Indistinct cleavage, conchoidal fracture, brittle to tough. Commonly 6-sided crystals, horizontally striated prisms, glassy grains. Colorless, rosy, smoky, brown, black, purple. Transparent to translucent, vitreous luster. Hardness 7 and specific gravity is 2.7. Infusible, insoluble,. In Pt loop fine powder with sodium carbonate fuses to clear glass BB. An abundant and ubiquitous mineral in sedimentary rocks, metamorphic and lighter colored igneous rocks. Varieties: rose quartz, smoky quartz, amethyst, milky quartz, citrine, aventurine, cat's eye, and tiger's eye.

The mineral group of composition SiO_2 is subdivided into the crystalline variety called quartz and the non-crystalline varieties of the same chemical composition. Each of these two major groupings include several mineral types dependent on color and form.

For the crystalline quartz varieties here described we illustrate several types. Color and/or form are the determinants for the specific name. Smoky quartz is as described. Amethyst, the purple variety, is well known for its peculiar, but beautiful color tone. Rose quartz, the pale red to pinkish variety, is a fairly common massive variety of quartz. Other types include milky or white quartz, citrine or yellowish quartz. Based on form and structure, there are crystal quartz, rutilated quartz (rutile needles throughout the mineral), ghost quartz (growth stages visible in the mineral), aventurine (mica, hematite, or other mineral-spangled quartz), cat's eye – an opalescent variety of quartz due to fibrous mineral inclusions, and tiger's eye – silica pseudomorph after crocidolite.

QUARTZ

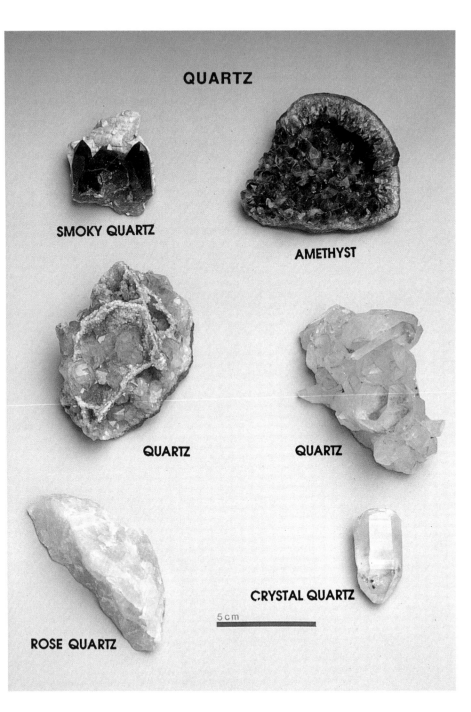

SMOKY QUARTZ

AMETHYST

QUARTZ

QUARTZ

CRYSTAL QUARTZ

5 cm

ROSE QUARTZ

OPAL   **Semi-precious Mineral**   **Non-crystalline**
SiO_2nH_2O **Varieties: Wood Opal, Fire Opal**
                    Tripolite, Diatomite, Siliceous Sinter, Hyalite

No cleavage, conchoidal fracture, brittle. Non-crystalline, massive. Colorless, white, red, green, blue, gray, brown, yellow, normally in pale shades. May have a rich play of colors, and on turning show almost an iridescence. Transparent to translucent and vitreous to resinous and waxy luster. Hardness 5.5-6.5 and specific gravity 1.9-2.3. Infusible. BB turns opaque and may decrepitate. Yields water in closed tube when strongly heated. Common in veinlets, cavities, may replace fossil wood.

Several varieties of opal are recognized. Wood opal is that formed by the replacement of wood. Fire opal has internal fire-like reflections of yellow, red and green like flashes of fire when turned in the light. Mexico and Australia are prime areas. It is prized as a semi-precious stone. Noble opal, the milky, almost opaque variety with a play of brilliant red, green, orange and other hues, occurs in Mexico, Honduras and Hungary. Tripolite is a variety of opal made up of the siliceous microscopic skeletons of diatoms of intricate form and shapes. The rock name for this accumulation of skeletal remains is diatomite. Siliceous sinter is a porous to dense, translucent to opaque, white, gray to brownish incrustation derived from hot spring water carrying silica in solution. Hyalite is clear, colorless and of globular, botryoidal and crust-like form and may be translucent. It has been called Muller's glass.

CRYPTO-   **Some Varieties as**   **Hexagonal**
CRYSTALLINE QUARTZ   **Gemstones**
SiO_2

Indistinct cleavage, conchoidal fracture, brittle to tough. No crystals, massive, botryoidal, banded, layered. White to gray usually, almost all other colors. Transparent to opaque. Vitreous to greasy luster. Hardness is 7 and specific gravity 2.5-2.8. Infusible, insoluble. In Pt loop mixed with half sodium carbonate fuses to clear glass BB. Abundant, ubiquitous, as cherts, chalcedony in veins and openings. Varieties: chalcedony, sard (brown chalcedony), Carnelian (red), chrysoprase (apple green), onyx (banded colors), agate (curved banded colors), moss-like forms, bloodstone or heliotrope (small red spots in green chalcedony), flint (opaque, dull, dark, white coating), chert (dull, opaque, light to dark, splintery fracture), jasper (dull, opaque, red, brown, yellow), prase (dull green, translucent).

OPAL

ONYX

BANDED CHALCEDONY

FLINT

OPAL

JASPER

FIRE OPAL

5 cm

CHALCEDONY with
WHITE QUARTZ

SANIDINE	**High-temperature Form**	**Monoclinic**
$(K,Na)AlSi_3O_8$	**of Orthoclase**	

Good cleavage, occurs as glassy and transparent, Colorless to white, tabular and square crystals in rhyolites, trachytes, phonolites and other flow rocks. Hardness is 6 and specific gravity 2.6. See orthoclase.

ORTHOCLASE	**Common Rock-forming**	**Monoclinic**
$KAlSi_3O_8$	**Silicate**	
MICROCLINE		**Triclinic**
$KAlSi_3O_8$		

Other:
Amazonstone – green microcline
Sanidine – glassy orthoclase (see above)
     Tabular, square crystals
Adularia – orthoclase (low temperature form)
     Colorless, transparent mainly prismatic crystals
Moonstone – adularia with pearly or bluish sheen

Two good cleavages at right angles, with uneven to conchoidal fracture, and brittle. Crystals usually prismatic, twinned, shapeless and massive. Colorless, white, light gray, cream, green, red, pink, red tone. Translucent to transparent with vitreous luster. Hardness is 6 and specific gravity 2.6. Fuses at 5 and is insoluble in acids. One of the most abundant of rock-forming minerals in igneous rocks, metamorphic rocks, some in veins. Common associates are quartz and muscovite.

PLAGIOCLASE	**Common Rock-forming Silicate**	**Triclinic**
$NaAlSi_3O_8 - CaAl_2Si_2O_8$		

     Isomorphous series
Albite –oligoclase-andesine
     labradorite-bytownite-anorthite
Moonstone – an albite with pearly-bluish sheen
Sunstone – oligoclase with scales which
     reflect in reddish and golden tones

Two very good cleavages (86° and 94°), conchoidal to uneven fracture and brittle. Tabular crystals, twins, massive, shapeless, striae on cleavages. Gray, white, also colorless, green, yellowish or reddish. Color plays often in labradorite and andesine. Vitreous luster and transparent to translucent. Hardness is 6.0-6.5 and specific gravity 2.6-2.8. Fuses at 4.0-4.5 to colorless glass. Albite insoluble in acid, while anorthite is decomposed by HCl. Abundant minerals in most intermediate colored igneous rocks, and some metamorphic rocks such as gneisses.

ORTHOCLASE

SANADINE

PLAGIOCLASE (Albite)

ORTHOCLASE

MICROLINE

5 cm

PLAGIOCLASE (Labradorite)

LEUCITE   Pseudo-cubic
$KAlSi_2O_6$

Indistinct cleavage, conchoidal fracture, brittle. Trapezohedral crystals, commonly striated. White to pale gray, vitreous to dull luster, translucent. Hardness 5.5-6.0 and specific gravity 2.5. Infusible. Moistened with cobalt nitrate solution and heated intensely BB turns blue (Al). HCl decomposes it. Rare, occurs in igneous rocks, silica poor.

NEPHELINE   Hexagonal
$(Na,K)(Al_3Si)_2O_4$

Distinct cleavage, prismatic, conchoidal fracture, brittle. Short prismatic crystals, 6- or 12-sided rare, also massive or granular. Colorless, white, gray,. rarely reddish to greenish. Of vitreous to greasy luster, and translucent. Hardness is 5.5-6.0 and specific gravity 2.6. Fuses at 3.5-4.0 to colorless glass, yields yellow (Na) flame. Powder dissolves in concentrated HCl, evaporate latter almost to dryness and silica gel forms in test tube. Uncommon, occurs in igneous rocks low in silica. Associates are feldspar, biotite, zircon, and sodalite, no quartz.

SODALITE   Silicate Mineral   Isometric
$Na_4Al_3Si_3O_{12}Cl$
Cancrinite

Distinct six direction cleavage at 60° angles, conchoidal to uneven fracture and brittle. Rare dodecohedral crystals, commonly massive, granular. Commonly azure blue, also violet, light gray, white greenish. *Hackmanite* is pink variety which fluoresces reddish. Transparent to translucent and of vitreous luster. Hardness is 5.5-6.0 and specific gravity is 2.1-2.3. Fuses at 3 to 5 to colorless glass. HNO_3 dissolves mineral, add silver nitrate and white precipitate occurs (Cl). Uncommon, in silica poor rock with nepheline and cancrinite. The latter is hexagonal, with prismatic cleavage, hardness 5-6, and specific gravity of 2.4-2.5. Color gray, greenish, yellowish, bluish and flesh red. Streak is white, luster vitrous and transparent to translucent. Common associate with Sodalite.

LAZURITE   Gemstone and Decorative Mineral   Isometric
$(NaCaK)_4Al_3(AlSi)_3Si_6O_{24}(ClFOHCO_3SO_4)$

Imperfect cleavage, uneven fracture, brittle. Rare crystals, cubes, octahedra, commonly massive. Blue to greenish blue. Streak is bright blue. Translucent to opaque, vitreous luster. Hardness is 5.0-5.5 and specific gravity 2.4-2.5. Fuses 3.0-3.5 Retains color upon cooling. Dissolves in HCl, produces H_2S gas. Green glow BB. Rare mineral in metamorphosed limestone.

173

NATROLITE Member of Zeolite Family Monoclinic
$Na_2Al_2Si_32H_2O$

Perfect prismatic cleavage, uneven fracture and brittle. Acicular crystals, usually radiating or divergent. Colorless or white, sometimes grayish or yellowish. Vitreous luster, transparent to translucent. Hardness is 5.0-5.5 and specific gravity 2.2. Fuses at 2 to colorless glass. Has yellow (Na) flame and yields water in closed tube. Soluble in HCl. Under UV light often fluorescent. Uncommon secondary mineral in cavities in basalt or other basic rocks. Calcite, other zeolites associates.

CHABAZITE Zeolite Family Hexagonal
$(KCa\ Na_2)\ Al(SiO_3)_4\ 6H_2O$

Distinct rhombohedral cleavage, three directions, not at right angles, uneven fracture. Cube-like crystals, twins common. Usually white, yellow, often purplish or red. Has vitreous luster and is transparent to translucent. Hardness 4-5 and brittle with specific gravity of 2.1-2.2. BB swells and fuses at 3 to bubbly glass. Yields water in closed tube, and is decomposed by HCl. Common zeolite and occurs in amygdaloidal openings in basalt with quartz, calcite.

ANALCITE Zeolite Mineral Isometric
$NaAlSi_2O_6\ H_2O$

Traces of cubic cleavage, but generally none, uneven fracture. Cubic, complex form, also massive. Commonly crystalline, generally trapezohedrons; also granular. Colorless or white; vitreous luster. Transparent to opaque. Hardness 5.0-5.5, brittle; specific gravity 2.2-2.3. BB fuses at 3 to clear glass, yields yellow flame (Na); in closed tube gives water; decomposed by HCl. Common mineral of zeolite family found in openings in igneous rocks, generally secondary in origin. Associates are other zeolites, calcite, apophyllite, prehnite, datolite.

HEULANDITE Member of Heulandite Group Monoclinic
$(Ca,Na_2)OAl_2O_3\ 6SiO_2\ 5H_2O$ (Mordenite to Brewsterite)

Perfect one-direction cleavage, uneven fracture and brittle. White, red, gray and brown. Transparent to translucent with vitreous to pearly luster. Hardness 3.5-4.0 and specific gravity of 2.2. Swells BB and forms worm-like mass. Fuses at 2.0-2.5 to white enamel. A zeolitic mineral.

STILBITE Silicates Monoclinic
$H_4(CaNa_2)Al_2(SiO_3)_6\ 4H_2O$

One perfect direction cleavage, uneven fracture and brittle. Sheaf-like aggregates of twinned crystals. White, yellow, brown or reddish. Transparent to translucent and of vitreous luster. Hardness 3.5-4.0 and specific gravity 2.1-2.2. Decomposed by HCl. Fuses 2.0-2.5 to white glass, swells. Yields water in closed tube. Uncommon, secondary mineral in cavities and in other basic igneous rocks. Associates are calcite and other zeolites.

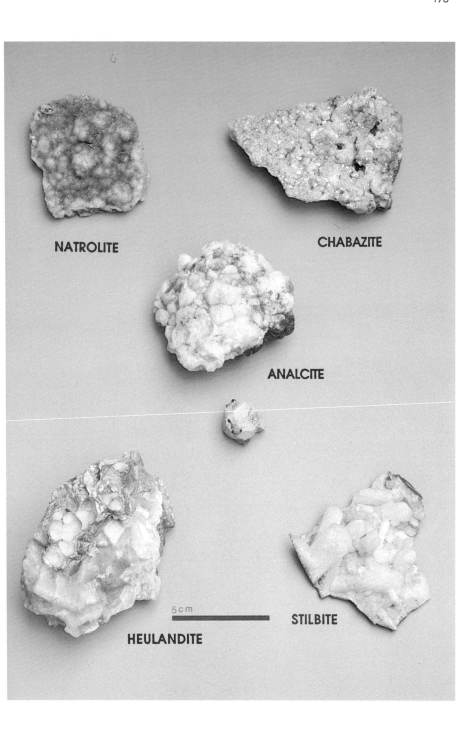

NATROLITE

CHABAZITE

ANALCITE

5cm

HEULANDITE

STILBITE

ROCK DESCRIPTIONS

Igneous Rocks

Metamorphic Rocks

Sedimentary Rocks

Extraterrestrial Rocks

Thin Sections (10X) of Rocks

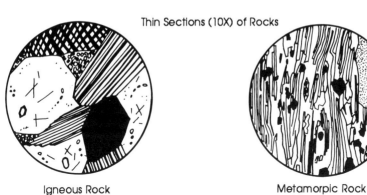

Igneous Rock

Metamorpic Rock

Sedimentary Rock

INTRODUCTION

Now that you have mastered the techniques of mineral identification, it is an easy additional step to identify the many rocks you will encounter in the field. Rocks, as you know, are the major building blocks of the continents and the ocean basins. Some rocks may be made up of a single mineral aggregate such as chromite which is composed solely of the mineral chromite, or rock salt of the mineral halite, or even ice of the mineral ice making up large glaciers on the surface of the earth. Other rocks consist of mixed mineral aggregates, or of fragments of former rock or mineral masses such as a granite or conglomerate.

Igneous, metamorphic and sedimentary rocks comprise the three major classes of rocks. Igneous rocks are derived from the cooling, crystallization and solidification of once molten matter generated within the earth and called a magma. Metamorphic (meta = change, morph = form) rocks form as a result of the action of heat, pressure or a combination of the two and hot fluids on a former rock. The rock mass changes to a more stable form under these environmental conditions. Metamorphic rock may be derived from an igneous, sedimentary or even an earlier formed metamorphic rock. Clastic sedimentary rocks form from the deposition of fragments or detritus of earlier formed rocks as a result of erosion, transportation, deposition, induration, and eventual consolidation to a rock. Chemical sedimentary rocks form from the results of weathering, solution, and transportation of chemical products and their eventual precipitation by chemical, biochemical or evaporative processes and later compaction and induration to form a rock mass. Much rarer rock types form from organic sedimentary accumulations to make peat, lignite, and coal through compaction, induration and some chemical changes. These interactions of rock forming processes are graphically described in the "rock cycle" (figure 27).

The more abundant minerals of igneous and sedimentary rocks are included in Table 14. These make up over 98 percent of the minerals in this part of the earth's crust.

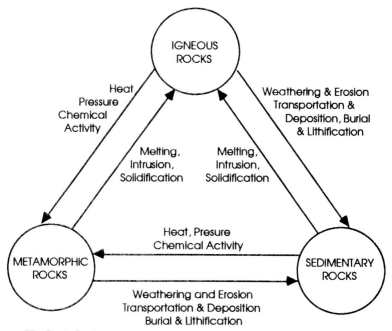

Figure 27. Rock Cycle

IGNEOUS

ROCKS

With Color Photographs

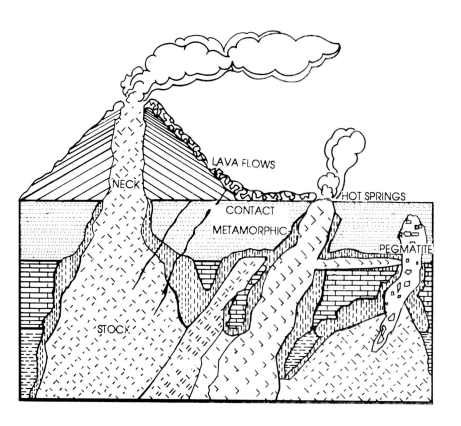

IGNEOUS ROCKS

Volcanic Rocks – Volcanic rocks solidify from a lava (magma) on the earth's surface and form lava flows. The rock which results from the crystallization of a lava is determined by the original composition and its cooling history. In contrast to a lava flow, a buried or partly buried magma, under high vapor pressure or its solidified equivalent, may be explosively ejected into the air. The resultant broken fragments or breccia and fine particles fall through the atmosphere and form ash deposits such as *tuff*, or ash and fragmental material called *tuff breccia*. In certain types of volcanic eruptions, partly solidified magma with included gases and molten magma may blow out almost horizontally from a vent and form glowing avalanches or *ash-flows*. Some of these eruptions which are very violent in character may cover tens of thousands of square miles in a series of eruptions. In the rare case where a magma, or lava or ash flow, arrives on the surface and is very rapidly cooled, a *glass*, or *vitrophere* (if crystals are visible in the glass matrix) may result. If very rapid loss of vapor occurs from a magma and it also cools quickly, a frothy, glassy rock results and *pumice* forms. This may also be blown explosively into the atmosphere and fall to the earth and be incorporated into other volcanic material to form *pumice tuff breccias*, or if the explosively generative particles are sand size or less, a *pumicite* may form.

TABLE 14. The More Abundant Minerals of Igneous and Sedimentary Rocks

Mineral	Igneous Rocks	Sedimentary Rocks
Feldspar (all varieties)	50%	16%
Quartz	21%	35%
Pyroxene, Amphibole, Olivine	17%	–
Micas	8%	15%
Magnetite	3%	–
Titanite, Ilmentite	1%	–
Clay Minerals	–	9%
Dolomite	–	9%
Chlorite	–	5%
Calcite	–	4%
Limonite, Hematite	–	4%
Others*	–	3%
Totals	100%	100%

*Others include: Leucite, Nepheline, Apatite, Sphene, Topaz, Zircon, and other rarer minerals.

More often, the magma does not reach the surface of the earth. In this environment it cools or solidifies below the surface in the host rock it has invaded or partly assimilated. Because of its much lower rate of heat loss in this setting, the crystallization process has more time, nucleation of crystals begin, seed minerals are able to grow larger, and the final solidified rock mass has a crystalline aggregate of minerals readily visible to the eye in a much finer-grained matrix. With longer cooling time, and possibly greater depth of the cooling magma below the surface, the mineral growth may finally be almost equigranular, medium to coarse-grained, and of phaneritic texture. These mineral crystals are called *phenocrysts* and the texture of the igneous rock is *porphyritic*. Thus, a rhyolite porphyry is a rock whose composition is very much like granite, but whose visible minerals are few to as much as 25% of the total rock volume. The minerals include quartz and feldspar, the latter mainly of the orthoclase variety.

The general color of the solidified igneous rock is not only a guide to its name, but also suggests the chemical composition, and in turn the minerals of the rock. The light-colored or felsic rocks are made up of an abundance of quartz and feldspars, either megascopic (visible) or microscopic in size. Rhyolites and light-colored pumice and granite are in this group. The darker colored, more iron-rich, or mafic (magnesium-ferric rich) igneous rocks consist of the more calcium-rich feldspars and the magnesium-iron bearing biotite, amphiboles, pyroxenes, and olivine. These include basalts and the depth equivalent gabbroic rocks of the crust. An intermediate colored and intermediate compositional group lies between the felsic and mafic groups. It includes the diorites and andesites, which are among the very abundant igneous rock types in the world.

The most simple classification of igneous rocks would be to describe them as light-colored, medium-colored and dark-colored. In essence we do this by listing them as felsic, intermediate, and mafic, and even ultra-mafic. The classification thus denotes feldspar-silica-rich rocks, through those which are made up of increasing contents of magnesium, calcium, and iron to those which consist essentially of high content of iron-magnesium-calcium with low silica content. The textural characteristic of these rock groups indicate the crystallization history of the rock.

The classification chart of the igneous rocks (Table 15) includes the more important rock types. It is based on texture, or grain size of the included minerals, color, and mineral composition. If the visible minerals can be identified in the field, the naming of the rock is a relatively simple matter.

Obviously, more refined methods are used to give an exact classification name to a rock type. This involves mounting a rock slice on a glass slide, grinding it to a very thin thickness until it is transparent, then studying the minerals and texture under a petrographic microscope, identifying them, and estimating their percentage, eventually arriving at a classification of the rock. New instrumental methods now permit rapid determination of the chemical composition of a rock at a reasonable cost.

The chemical information is also used to classify the rock. Yet, even with these sophisticated methods, the field classification of an igneous rock can, if carefully done, yield an accurate mineral and chemical composition for the rock type and a specific name.

In summary, to classify an igneous rock first identify the visible minerals, note the color of the rock, refer to the general position of this color on the rock chart (table 15), identify the column for the visible mineral composition, then find the row which best matches the texture of the rock. The chart block lists the rock name. As an example of the procedure, if the rock is light colored, has quartz and feldspar and minor biotite in a matrix of finer-grained or aphanitic material, and the phenocrysts comprise about 20 percent of the volume, the rock would be a rhyolite porphyry.

Photographs of the common igneous rock types are shown in the accompanying color plates. Examples include rhyolite, rhyolite porphyry, andesite porphyry, granite, diorite, and several others. Pegmatite is also classified as a rock type. It has very large mineral sizes compared to the average igneous rock.

TABLE 15. Igneous Rock Chart

Texture	Light Color		Intermediate Color	Dark Color
Fragmental	Pumice, Tuff and Tuff Breccia			
Glassy & Phenocrysts	Obsidian – Tachylyte Vitrophyres			
Aphanitic Rare Pheno-crysts in very fine-grained matrix	Trachyte	Rhyolite	Andesite	Basalt (May be smooth vesicular, ropy amygdaloidal or breccia)
Porphyritic Fine-grained Matrix	Trachyte Porphyry	Rhyolite Porphyry	Andesite Porphyry	Basalt Porphyry
Phaneritic somewhat equigranular	Syenite	Granite and Variants	Diorite	Gabbro Pyroxinite Hornblendite Peridotite Dunite Olivine-Pyroxene
	Pegmatite (Very Coarse-grained)			
	Essential Visible Minerals			
	Feldspar Some Dark Minerals Occ. Nepheline	Quartz Feldspar Minor Dark Minerals	Feldspar (with striae) 20%-40% Dark Minerals	40%-70% Dark Minerals Biotite Hornblende Pyroxene Olivine Magnetite

Glasses

Red and Black Obsidian: The volcanic glass with a bright vitreous luster, and very distinctive conchoidal fracture is known as obsidian. Its specific gravity is about 2.4. The volcanic glass is the result of sudden cooling or quenching of the original hot magma. High viscosity of the magma may also have prevented crystal growth with the result of a glass on solidification. In either case, time from the molten to solid stage was insufficient to permit crystal growth and glass was the final product. The reddish coloration is probably the result of hematitic iron content. The blackness may relate to very fine-grained magnetite spread throughout the obsidian. The water content of glasses is usually less than 1 percent in obsidian, some 3-4 percent in perlite, and 4.10 percent in pitchstone. Specific gravity ranges from 2.37 for rhyolitic glass to 2.77 for basaltic glass. Most natural glasses have a composition approaching the quartz-alkali feldspar cotectic.

The black variety is more typical of the volcanic glasses, but both the red and black types are characterized by well-developed conchoidal fracture. This characteristic and the very sharp fracture edges which form were features sought by the American Indian for good chipping stone for arrow and spear heads. Currently, surgical blades are also made from high-quality obsidian because of the extremely sharp and clean edge formed when the rock is fractured. This rock type is most generally of extrusive origin and of felsic composition.

Vitrophere: An obsidian which has conspicuous phenocrysts in its glassy matrix is called a vitrophere after *vitro* = glassy and *phyre* = crystals or phenocrysts. The phenocrysts apparently formed at depths below the earth's surface while the magma was still molten. The latter was then extruded and the earlier formed phenocrysts were frozen into the glassy matrix which formed upon rapid cooling and solidification. Phenocrysts seldom exceed an eighth of an inch in size. Colors of vitropheres may be black, red, or dark tones of green and greenish blue. Common vitrophere micro-phenocrysts are usually magnetite, ilmenite and other oxides. Phenocrysts of feldspar are not uncommon.

Spherulitic Glass: In some obsidians fluids are trapped in the rapidly cooling magma. These tend to form a void or open space from the high vapor pressure exerted by the fluid. Sublimates form from these fluids on cooling and microcrystalline structures develop. These form spherulitic masses within the glassy matrix and are referred to as *lithophysae*. Another popularized term for the special black obsidian with white spherulites is *Apache Tears*. The spherulites consist of high temperature quartz of the varieties tridymite and cristobalite.

RED OBSIDIAN

BLACK OBSIDIAN

VITROPHERE

SPHERULITIC GLASS

5 cm

Glasses and Breccias

Pumice: A light colored rock, generally white to grayish brown, with a frothy texture. Under the hand lens fine fibers surround a multitude of open pores and vesicles. This high pore space imparts a low specific gravity to the volume of rock; indeed, it may even float on water. The composition approaches that of rhyolite, but some is basaltic. The rock and its unique texture result from the rapid escape of gas in a near surface or surface emplacement environment, with extreme vesiculation and rapid cooling of the magma during the degassing period. Rock type is of extrusive origin, and usually felsic in composition.

Pumicite: A light-colored, glassy rock made up of fine glass particles which are products of explosive activity of a very viscous magma. The composition is generally rhyolitic. The rock may be layered as a result of air fall of the particles which were violently exploded into the atmosphere as finely divided glass fragments and semi-molten material. The extremely fine-grained glassy aggregate is used as an abrasive in various polishing agents including car polishes, tooth paste and other products.

Welded Tuff Breccia: This rock is light gray to moderately dark red in color, has angular lithic or rock fragments included in a dense matrix with broken phenocrysts. The term "welded" refers to the dense matrix of the rock which resulted from relatively rapid cooling of a formerly molten mass. Welded tuffs and tuff-breccias form from explosive ejecta from a main venting area in which solids, semi-molten magma, gas, and water are violently erupted as almost horizontally travelling masses with velocities up to and more than 100 miles per hour. In almost instantaneous geologic time areas measured in thousands of square miles are covered by these "glowing cloud" masses. Compaction of the rock mass, escape of volatiles and rapid cooling bring about the welded character of the rock unit. In some cases even columnar structures to scores of feet in length may develop in these welded tuffs.

Tuff Breccias: With light to dark colored, but characteristic angular rock fragments in a fragmental matrix, tuff-breccias resemble the welded tuff breccia, but lack the dense, more solid matrix. The tuff breccias are quite friable, may powder easily, and have a lower specific gravity. Fragment size may range to inches in diameter and may be a function of distance from the explosive center.

PUMICE

PUMICITE

TUFF BRECCIA

5 cm

WELDED TUFF BRECCIA

Light-colored Felsic Igneous Rocks

Rhyolite-Topaz bearing and Rhyolite: Generally light in color in shades of gray, but sometimes red, rhyolites have a dense matrix and minor phenocrysts of feldspar and quartz within them. In some cases there is distinctive and conspicuous banding or layering, often times intricately folded. This is the result of the flow of the original lava and planar development in the cooling molten mass. Gas, under pressure, may develop pore space within some rhyolites and, in turn, minerals may crystallize within these. Topaz crystals, some to 1 inch and more in length are a common feature in the topaz-bearing rhyolite at Topaz Mountain, Utah. Silica content is in excess of 72%, Al_2O_3 about 13.5%, Na_2+K_2O some 8% and CaO 1%. These are extrusive felsic igneous rock. Major minerals are quartz, sanadine and minor minerals are glass, biotite, albite, some magnetite, ilmenite. Rarer minerals include topaz, amphibole, diopside.

Aplite: Generally grayish to pinkish in color, sometimes medium red, and with a distinctive sugary texture, aplites occur as dike-like bodies cutting granites and country rock. The dikes range from inches to feet in thickness and may extend to thousands of feet in length. Composition is mainly quartz, potassium feldspar and some plagioclase. Rock type is intrusive (low depth) felsic igneous rock. Major minerals are quartz, potassium feldspars, mica. The minor minerals include tourmaline, plagioclase and others.

Granite: The texture is characterized by equigranularity, color of light to intermediate tones, generally in the grays to reds, and a uniform mix of minerals making up the rock. Granite is a common rock in the crust of the earth, having formed at depths where crystal growth, because of slow-cooling, continued until mutual interlocking boundaries formed. With a high silica composition, quartz, orthoclase feldspar, lesser plagioclase, and even lesser ferromagnesian minerals compose the rock. Biotite is common and the term biotite-granite is applied, but hornblende, too, black and elongate, may be the dark mineral and the modifier hornblende granite is used. Silica content exceeds 70% for the normal granite. The rock type is intrusive felsic igneous rock. Major minerals are quartz and potassium feldspar. Minor minerals include plagioclase, mica, magnetite and ilmentite.

Pegmatites: The igneous rock characterized by very large grain size, some minerals inches to feet in length, is called a pegmatite. The most common is of granitic composition, but is also known for gabbro. The granite variety approaches granite in composition but has the very large crystal size of the pegmatites. Simple pegmatites have a granite composition. Layered pegmatites, a layering mainly of granitic composition, and zoned pegmatites, a concentric pattern with the core carrying unique assemblages of rare minerals including large spodumene, amblygonite, columbite-tantalite, and other rare earth minerals. The major minearls of pegmatites can reach incredible crystal dimensions. Spodumene and beryl crystals of several feet in size were mined in pegmatites in South Dakota and Colorado. The rock type is intrusive (low depth) and usually of felsic composition. Major minerals are quartz, potassium feldspar, mica. Minor minerals include tourmaline, beryl, zircon, apatite, and other rare minerals mentioned above.

RHYOLITE-TOPAZ BEARING

RHYOLITE

APLITE

GRANITE

BIOTITE GRANITE

PEGMATITE

5 cm

Hornblende Phonolite: Light to medium gray, dense matrix, occasional phenocrysts of feldspar and felspathoids (nepheline, leucite, and others) and minor ferromagnesian minerals (usually hornblende) are characteristic of the hornblende phonolite. The rock has an unusual ring to it when struck by a hammer. The rock is silica deficient and chemically shows only 57 percent of silica in contrast to a rhyolite with 73 percent. The rock occurs as lava flows and dikes. Potassium + sodium oxide content is about 14% and alumina near 21%. This is an intermediate composition extrusive igneous rock. Major minerals are sanidine, nepheline, acmite and hornblende (riebeckite or artvedsonite). Minor minerals include anorthoclase, albite, augite, apatite, titanite, magnetite, sodalite, analcite, zeolites and melanite garnets.

Leucite Porphyry: The distinctive dodecohedral crystal form and the light to medium gray dense matrix are characteristics of leucite porphyry. Ferromagnesian minerals are rare.

Growth of the large phenocrysts must have taken place in a slow-cooling environment with the early leucite growing from nucleation centers and achieving an almost perfect crystal shape. The isometric form is readily recognized. The magma then moved upward to the surface environment where the dense matrix formed from much more rapid cooling. Silica content approximates 58% and potassium oxide approaches 6 percent. This is an intermediate extrusive igneous rock whose major mineral composition is leucite. Occasional magnetite is present.

Nepheline Syenite: The depth equivalent of the surface phonolite is expressed in nepheline syenite, an equigranular, light to intermediate gray rock with nepheline, orthoclase feldspar and minor accessory ferromagnesian minerals. The rock apparently developed in a relatively slow cooling environment and the various minerals were able to grow with mutually interlocking boundaries. The rock contains about 55 percent silica, and relatively high content of potassium and sodium oxides, 14%, and about 20% Al_2O_3. The rock is an intermediate intrusive igneous rock. Major minerals are potassic feldspar, plagioclase, amphibole, nepheline. Minor minerals are pyroxene, quartz, biotite, magnetite, ilmenite. Occasional minerals include olivine and corundum.

189

HORNBLENDE PHONOLITE

LEUCITE PORPHYRY

NEPHELINE SYENITE

5 cm

Andesite Porphyry: While phenocrysts are not abundant in this intermediate gray porphyritic rock, they are mainly of plagioclase. The mafic mineral content is relatively low and the matrix is dense but of darker gray tone. The rock is part of a lava flow.

The usual andesite contains more mafic minerals, usually hornblende with more or less biotite and/or augite. Quartz is less than 10 percent if present. Silica content averages 60 percent, and calcium content slightly exceeds potassium and sodium content combined. This is an intermediate extrusive igneous rock. Major minerals include plagioclase (anorthite <59%), biotite. Minor minerals are magnetite, ilmenite, quartz, hornblende, pyroxene. Occasional minerals are orthoclase, anorthoclase and olivine.

Latite Porphyry: Conspicuous phenocrysts of plagioclase feldspar of two generations, a larger and small size, with orthoclase feldspar intermixed in a dense gray matrix characterize the latite porphyry. A few grains of biotite are visible, quartz is not. The rock is part of a lava flow. Silica content of latite porphyry approaches 56 percent and potassium + sodium oxide content is in excess of calcium oxide on the order of 8 percent to about 6 percent. This is an intermediate extrusive igneous rock. Major minerals are sanidine, plagioclase, augite and hornblende while minor minerals include magnetite, ilmenite, anorthoclase, olivine and feldspathoids.

Diorite: The fine-grained texture and dark gray color are common to diorite. The rock is holocrystalline (all minerals are crystalline and visible to the unaided eye) and the light to dark mineral content is about 65%/35%. Hornblende is the dark mineral. Silica content averages 57 percent and the potash and soda to calcium oxide is about 5.5/6.7 percent respectively. Density averages 2.85. This intermediate intrusive igneous rock has major minerals which include plagioclase (anorthite < 50%) and hornblende. Minor minerals are quartz, allanite and ilmenite. Occasional minerals are pyroxene, K-feldspar and biotite.

Monzonite: An intermediate gray tone, an equal abundance of both orthoclase and plagioclase feldspar and about 15% ferromagnesian mineral content in the form of hornblende characterize this equigranular monzonite. Quartz is not visible. Silica content for a monzonite is about 56 percent and soda plus potash/calcium oxide is 6.4 percent to 6.5 percent respectively for the normal monzonite. Density aproximates 2.82. This intermediaet intrusive igneous rock has K-feldspar, labradorite, pyroxene and hornblende. Minor minerals include quartz, biotite, titanite, magnetite and ilmenite. Less common minerals are olivine and nepheline.

ANDESITE PORPHYRY

LATITE PORPHYRY

DIORITE

MONZONITE

5 cm

Dark-colored Mafic Igneous Rocks

Scoriaceous Basalt: This rock has an unusual scalloped structure; irregular shapes and ragged textures surround ovoid openings. The rock matrix is dense and few or no phenocrysts are visible in the dark colored rock. The openings are the result of vesiculation by the rapid escape of large quantities of gas during the rapid cooling of a large lava flow. The rock is a scoriaceous basalt, a term commonly used to indicate a basaltic pumice in contrast to the light rhyolitic pumice earlier noted. Silica content approximates 50 percent and calcium oxide about 10 percent for the basalt. Lesser content of potash and soda is shown by the less than 3 percent total. This is an extrusive mafic igneous rock whose major minerals are plagioclase (anorthite >50%) and pyroxene. The minor minerals are glass, olivine, magnetite, apatite and quartz.

Basalt: A few ferromagnesian phenocrysts, mainly augite, in a dense and dark colored matrix, characterize this basalt. The increased density over the lighter colored igneous rocks is apparent.

Normally basalt consists of dark gray plagioclase feldspar in a dark-colored matrix. Augite and olivine are commonly present, while hornblende and biotite are scarce. Some magnetite may be visible. Silica content approaches 50 percent for the basalt.

Basalt Porphyry: A fine-grained matrix and small phenocrysts of probable augite and minor plagioclase feldspar within it characterize this dense basalt porphyry. The feldspar present is dark gray and the darker tone of the rock tends to obscure it. Silica content approximates 50 percent, and calcium oxide is about 10 percent.

Gabbro: A dark gray color results from approximately 50% light and 50% dark mineral content which characterize this gabbro. The lighter-colored minerals are mainly plagioclase feldspar and the dark minerals mainly pyroxene. The rock is equigranular. Silica content approaches 48 percent, 11 percent calcium oxide and the $Na_2O + K_2O$ content is about 1/3 that of the calcium in the rock.

SCORIACEOUS BASALT

BASALT

BASALT PORPHYRY

GABBRO

5 cm

Ultra-Mafic Igneous Rocks

Anorthosite: Coarse-grained, interlocking plagioclase compose this intermediate gray igneous rock. Minor pyroxene is also visible between the feldspar grains. It is not a common rock. Ilmenite-magnetite bodies are commonly associated with it world wide. It also appears to occur mainly in a specific time window in Precambrian time about two billion years ago. The composition approaches that of the calcium rich anorthite and labradorite plagioclase feldspars.

Dunite: This rock is coarse to sugary-grained, pale green to brownish in color and almost mono-minerallic in composition with olivine. This dunite is an unusual early-formed rock. Chromite is commonly an accessory mineral and sometimes is in abundance. The rock may represent a segregated fraction of a differentiated magmatic intrusion. Silica content and magnesium and iron contents are those of its component mineral olivine.

Hornblendite: Black hornblende, coarse-grained, interlocked and denseness characterize this ultra-mafic hornblendite. It, too, must represent a segregated fraction of an ultra-mafic magma. Composition-wise the rock approaches that of its main constituent, hornblende. It is not a common igneous rock.

Pyroxenite: The rock is dark-colored, dense, and is almost mono-minerallic in composition with pyroxenite as the main mineral. It is a probable fraction of a differentiated magma. The mineral making up the rock is coarse-grained and the composition that of the general pyroxene group.

Kimberlite: (No photo) This rock is the major source of diamonds. It is a potassic, porphyritic, serpentized basaltic alkalic or carbonate-rich mica peridotite, containing large phenocrysts of olivine, enstatite, Cr-diopside, phlogopite, garnet, carbonates and other minerals. Kimberlites form breciated diatremes and dikes and contain fragments and inclusions of limestone through ultramafic rocks to eclogites (biminerallic rock: garnet and jadeite and diopside end-members) dispersed in the ultramafic matrix. The magma was extremely high in volatiles before cooling. The rock occurs as clusters of pipes and dikes in a random geographic setting along major tension joints in and along rift valleys and fracture systems mostly in shield areas, cratons, rifted platform areas and plateaus.

For the rockhounds: don't get your hopes too high. Kimberlites are extremely hard to find and the diamonds within them even more so. Surprisingly though, Kimberlite masses, with diamonds, have recently been discovered in southern Wyoming not too far from the "Great Diamond Hoax" site of the last century. However, only 1 out of every 100 kimberlite pipes produce diamonds economically. A good diamond mine, like the Kimberly mine in South Africa, mines 1600 metric tons of rock to recover one carat (or 0.2 grams) of diamond.

ANORTHOSITE

DUNITE

HORNBLENDITE

PYROXENITE

5 cm

METAMORPHIC

ROCKS

With Color Photographs

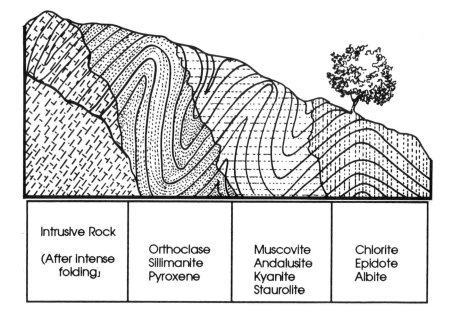

Intrusive Rock (After intense folding)	Orthoclase Sillimanite Pyroxene	Muscovite Andalusite Kyanite Staurolite	Chlorite Epidote Albite

Metamorphic Rocks

Have you observed what happens to various substances when heat is applied or when other components are mixed with the original substance and heat is applied? The material may change in form, it may become more fluid and tend to flow, or it may combine with the added components and form a new solid substance. In nature rocks behave in a somewhat similar fashion. Given a certain rock type, the addition of heat and or pressure of increasing intensity over a period of time plus the presence or addition of fluids to this environment causes the rock to adjust to these new conditions. A new equilibrium is established through the formation of more stable minerals. How much change in form occurs is dependent on the magnitude of the stress and temperature and the reaction of the fluid components, plus the duration of the reaction time. Simple rocks such as shale, a sedimentary rock mainly composed of clay minerals and some quartz, may be radically changed. The end product is a metamorphic rock, often times completely different in appearance. It may become a phyllite, a rock much hardened and with a visible sheen of micaceous minerals developed within it. Under stress the original shale may be converted to a slate. Where once the shale broke along bedding planes, now the rock parts along cleavage planes, often times completely different than the still visible bedding planes of the shale. Under the microscope one can observe the orientation of minute mineral grain parallel to this new breaking of cleavage surface. With even greater heat and stress the original shale may be changed to a schist, a metamorphic rock with clearly visible foliation planes and platy minerals aligned parallel to the schistose structure. Under extreme heat and pressure all of the components of the rock may be reorganized, even flowage may occur, and dark and light minerals tend to segregate together to form a banded-rock called a gneiss.

As in the case of the origin of the deeply buried igneous rocks, we have no way of observing these changes as they occur. We must base our conclusions on field observations and the inferred conditions of temperature and pressure. Fortunately, abundant and well-controlled laboratory experiments have been performed which permit well-documented conclusions as to the effects of temperature, pressure and fluids on various rock materials and the end products produced. The development of the ultra-high pressure tetrahedral press by Tracy Hall and his associates has resulted in productive research at pressure over 1,000,000 pounds per square inch and at temperatures up to 1500 degrees Centigrade.

Two stress environments for metamorphic rocks are recognized: 1) static or load metamorphism under elevated temperatures, and 2) dynamic stress environment with increased temperature. A third environment of metamorphism is that of increased temperature. The latter may act on the rock in both stress environments and locally only as heat near a cooling magmatic body.

With these factors operating in nature certain visible changes are wrought by metamorphism. The constituents of the rock may be completely reorganized to form a new more stable mineral assemblage. Grain size of the minerals is generally increased with a resultant coarser texture. The internal structure of the rock may change, and commonly a parallel arrangement of the constituent minerals may take place. In many cases the overall composition of the rock may remain the same, while in others constituents are lost or others added. The terms non-additive and additive metamorphism have been given to these processes.

Because the metamorphic processes take place at high pressure and high temperature, many of the minerals which form are also common to igneous rocks. Exceptions are the silica deficient feldspathoids such as leucite, nepheline and sodalite.

Table 16 includes those minerals common in metamorphic rocks. Descriptions of the individual minerals are included in the text.

Texture of Metamorphic Rocks: Recognition and naming of a metamorphic rock requires identification of the major minerals and the textures and structures present. The list of minerals includes most of those common to this rock class. The processes of metamorphism generally produce increase in mineral size, whether by recombination of elements and new mineral development, or by simple crystal growth of preferred minerals. Distinct differences in textures of igneous, sedimentary and metamorphic rocks exist. For most metamorphic rocks crystals are readily visible, and while somewhat interlocked form a mosaic often with rough to well-developed layering characteristics. Some minerals tend to develop good crystal form and are called idioblastic with bounding crystal faces. Xenoblastic crystals are those without crystal form. A porphyroblast is a crystal much larger than the matrix and may contain other mineral inclusions. It is of late stage development in contrast to phenocrysts of igneous rocks which are formed early in the crystallization process. Porphyroblasts may lie athwart the general internal structures of the rock or may show structures bending around them, as if deformed by the developing mineral.

Structures of Metamorphic Rock: Three types of structures are recognized in the metamorphic rock class: 1) planar-parallel planes or flat surfaces of minerals parallel or roughly so; 2) a series of linear or line-like features, such as elongate minerals parallel in their long direction which may or may not lie within the planar structure; 3) massive-linear or planar features not visible. One or more of these may be used in the naming of the metamorphic rock.

Classification and Naming of Metamorphic Rocks: Utilizing texture, structure, and mineral composition we can classify or name the metamorphic rock. In some cases the name may be derived solely from the structural characteristics, such as schist. In others it may be based on the mineral composition, as for quartzite or marble. Additional qualifying terms may give a more specific name to the rock. For a rock with schistose structure and clearly visible biotite, garnet and quartz present, the name would be Garnet biotite quartz schist based on the relative abundance of each of the mineral phases present, the least abundant first. The panacea of a name which would describe origin, original rock type and intensity of metamorphism and resultant mineralogy, texture and structure has not been derived. We must depend on visible minerals, textures and structures which may or may not denote the rock history.

The chart of Table 17 is general in its make-up but includes the main criteria for naming of this important rock class. It is based on typical structures, textures, and minerals.

TABLE 16 COMMON MINERALS IN METAMORPHIC ROCKS WITH MOST LIKELY ROCK HOSTS

Actinolite	S-M	Kyanite	S
Amphibole (hornblende)	S-G	Magnetite	CM-S-G
Andalusite	Hfls-S	Muscovite	Sl-Phy-S
Biotite	S-G	Olivine	M-Eclogite
Calcite	M	Phlogopite	M-S
Chlorite	Sl-Phy-S	Pyrite	widespread
Diopside	M	Quartz	S-G-Q
Dolomite	M	Serpentine	Alt. mafic rocks M-CM
Epidote	S-M	Sillimanite	S
Feldspar varieties	G-S-CM	Staurolite	S
Garnet (varieties)	S-CM	Talc	S
Graphite	Sl-Phy-S-M	Tremolite	S-M
Hematite	S-CM	Wollastonite	M
Hornblende	G-S		

*G=gneiss, S=schist, Q=quartzite, Sl=slate, Phy=phyllite, CM=contact metamorphic, M=marble, Hfl=hornfels

TABLE 17. CLASSIFICATION OF METAMORPHIC ROCKS

STRUCTURE	TEXTURE	MINERALS	ROCK NAME
MASSIVE TO FAINTLY LAYERED	Very fine grain	Identifiable under microscope	Hornfels
	Sugary texture, interlocking grains (Faint to distinct layering)	Mainly quartz	Quartzite
	Fine to coarse grains, with or without dark-streaking. Sometimes breciated.	Calcite, Dolomite	Marble
GRANOBLASTIC	Variable grain size	Silicates of calcium, iron, magnesium in or with limestone or dolomite, and near igneous contact. Includes garnets, epidote, pyroxene, amphibole and others.	Tactite
	Medium to coarse-grained rock, faintly foliated	Plagioclase, amphiboles, possible garnet, quartz, epidote	Amphibolite
		Feldspar, pyroxene, garnet, kyanite and other silicates	Granulite
FOLIATED	Very fine-grained, well foliated. Remnant sedimentary structures visible.	Mineral grains very small, microscopic	Slate
		Mica and quartz when recognizable	Phyllite
	Faint foliation to schistose and well-developed layering	Chlorites, plagioclase, epidote	Chlorite schist
		Muscovite, quartz, biotite	Mineral qualifier, as mica schist, biotite, muscovite, sericite schist, or sillimanite mica schist, kyanite, etc.
		Amphibole, plagioclase	Amphibole schist
	Gneissic-banded (dark-light bands)	Feldspar, amphibole, quartz, biotite	Gneiss
	Coarsely-banded mixture or metamorphic & igneous rocks	Feldspar, amphibole, quartz, biotite interlayered and interfingered	Migmatite

Foliated Metamorphic Rocks

Slate: This rock is homogeneous in its very fine-grained texture and composition, though it shows a difference in color banding. Schistosity, or cleavage, is independent of the bedding. The rock can be cleaved to relatively thin plates parallel to the finely developed schistosity. The latter is the result of the microscopic alignment of platy minerals such as mica and chlorite which have been oriented in the plane of cleavage as a result of stress. Numerous types of slates have been described, but the use of a color term such as gray slate or red slate dominates, though compositional terms such as graphitic and calcareous are also used.

Muscovite Schist: Crystalline, somewhat crenulated, mainly composed of grainy muscovite and showing distinct foliate structure are the characteristics of this muscovite schist of gray tone. It is the result of regional metamorphism of the original rock, probably a shale.
    The grain size for this rock is far larger than that of the slate and is clearly visible to the eye.

Garnet Mica Schist: Of lighter color tone, but with visible platy muscovite mica of foliate structure and containing small brown garnet metacrysts, this garnet mica schist is a distinctive and beautiful stone. It, too, is the result of regional metamorphism. Grain size is finer than the muscovite schist above, but clearly visible as is the foliate structure. Other mica schists may show large metacrysts of biotite, andalusite, staurolite and other minerals in a finer-grained matrix of mica and other minerals.

Amphibolite: A product of regional metamorphism this amphibolite is of higher-grade metamorphism than the schists. The main mineral is of the amphibole group, and is hornblende. Other minerals, such as plagioclase, are present but not readily visible. Minor pyrite is also present. A faint foliation is apparent on the sides of the rock specimen. The rock probably developed from the metamorphism of a basalt. The photographic view is downward onto the planar structure.

SLATE

MUSCOVITE SCHIST

GARNET MICA SCHIST

AMPHIBOLITE

5 cm

Thermally Metamorphosed Carbonate Rocks

Marble: The white rock is a contact metamorphosed limestone, that is, a rock subjected to increased heat near the contact of an igneous intrusive body. It shows increased grain size and a much lighter color than the original rock. Grains are almost equal in size, bedding planes have been essentially obliterated, and the rock considerably changed in appearance. The lack of other minerals within the marble suggests the original rock was a pure limestone.

Actinolite Quartz Tactite: A carbonate rock has been converted to an actinolite-quartz tactite by thermal additive metamorphism. The term "additive" refers to the addition of various elements by fluids from the nearby pluton which resulted in new mineral development in place of the carbonate rock. The elongate actinolite approximates one inch in length. Essentially all the original carbonate rock has been replaced by the silicate minerals.

Garnet Marble Tactite: An impure limestone in a contact zone near a pluton has been converted to this tactite or silicated marble rock. Possibly some introduction of silica, iron and aluminum took place to effect the development of the abundant garnet. It is not unusual to find scheelite, the $CaWO_4$, as a mineral component in these contact metamorphic zones. This type rock and its geologic seting was a prime target for geologists in World War II in their search for tungsten deposits.

Biotite Marble: The calcite of the original limestone has been recrystallized as a result of heat from a nearby intrusive mass. Grain size is coarse, crystal sizes almost equal, and the rock lightened in color to gray, possibly through partial loss of carbonaceous material originally present in the limestone. Bedding has been essentially destroyed by the thermal metamorphism. Small grains of biotite have developed from impurities within the original rock.

WHITE MARBLE

ACTINOLITE QUARTZ TACTITE

BIOTITE MARBLE

GARNET MARBLE TACTITE

5cm

Thermally Metamorphosed Siliceous Rock

Banded Hornfels: An original shale is here converted to an extremely fine-grained hornstone or hornfels through the action of heat from a nearby intrusive mass. The heat may have effected loss of water and possibly carbon dioxide, but the end product is a hard, fine-grained hornfel of about the same composition as the original shale. Bedding planes are preserved but parting along them is poorly retained.

Spotted Hornfels: An argillaceous rock with conspicuous spots up to almost an inch in diameter and very-fine grained is the end product of thermal metamorphism. The rock is almost flint-like in texture and very hard and brittle. The spots show some migration of components within the spot-like features and take on the appearance of "mothballs". These are localized in this rock unit within the thermally altered halo of a pluton.

Serpentinite: This green, almost wax-like stone, is a hydrated magnesium silicate of approximately $3MgO\ 2SiO_2\ 2H_2O$ composition. It is an unusual rock in that the mineral serpentine, of which it is composed, lacks regular crystal form but appears to be pseudomorphic or replacements of pre-existent minerals. The rock may have formed by replacement processes in ultra-mafic rocks and/or dolomites and lime-stones. In large masses serpentinite has been quarried as an ornamental stone.

Asbestos and Serpentinite: Serpentinite here, of the possible origin suggested above, is "veined" by a variety of asbestos called chrysotile. This fibrous mineral may have developed directly from the serpentinite or have been introduced by fluids moving along fractures in the rocks. The mineral has been mined in large quantities in Canada. It is used in brake linings and other products where high heat resistance is required.

BANDED HORNFELS

SPOTTED HORNFELS

SERPENTINITE

SERPENTINITE with CHRYSOTILE
(Asbestos)

5 cm

Regionally Metamorphosed Rocks

Quartzite Conglomerate: Pebbles or clasts of quartzite lie in a finer-grained matrix of quartz grains. The entire rock mass is of quartz composition, and in breaking, the fracture cuts through rather than around the pebbles in contrast to a normal conglomeratic rock. The welding of the grains and pebbles into an extremely coherent mass is probably the result of high heat and pressure on a quartz-rich sedimentary conglomerate. Relict bedding is still visible in the rock.

Banded Fine-grained Quartzite: Fine-grained sandstone has been converted to a fine-grained, banded quartzite. The mineral grains fracture when the rock is broken. Bedding is parallel to the color banding. A semblance of relict fine cross-bedding is visible on the left side of the rock specimen.

Micaceous Quartzite: An original clayey sandstone is metamorphosed to a micaceous quartzite. The micaceous (muscovite) layers lie between relatively pure quartzite. The thin-bedding parallels the page. Such rock is often used as a decorative stone in fireplaces and walls.

QUARTZITE CONGLOMERATE

BANDED FINE-GRAINED QUARTZITE

MICACEOUS
QUARTZITE

5 cm

Regionally Metamorphosed Rocks

Gneisses: The term gneiss applies to a group of rocks with banded or coarsely foliated structure. These are coarse grained, often irregularly banded and consist of mainly quartz and feldspar with some biotite. Some varieties occur near intrusive masses intricately interlayered in bedded metamorphic rocks and are known as migmatites. They are believed to have formed from the emanations from the partial melting and recrystallization of the original rock under high temperature conditions.

Other gneisses are formed from the conversion of sedimentary rock units into a more stable mass under the high temperature-pressure conditions and are called paragneisses. A third type, known as orthogneisses, develops from the metamorphism of primary igneous rocks.

Streaked Gneiss: The layering or foliation of the rock is distinct and made up of feldspar, quartz and biotite, but is streaky. Some individual layers end rather abruptly, others are quite continuous.

Folder or Contorted Gneiss: Well-developed bands of the gneiss are folded into relatively sharp folds a few centimeters or inches across. The bands consist of feldspar (white) and biotite-feldspar (gray).

Banded Gneiss: The two lower banded gneisses of the page have characteristics similar to the streaked gneiss. Both share discontinuity in the layering and compositions are about the same.

STREAKED GNEISS

CONTORTED GNEISS

BANDED GNEISS

5 cm

BANDED GNEISS

SEDIMENTARY

ROCKS

With Color Photographs

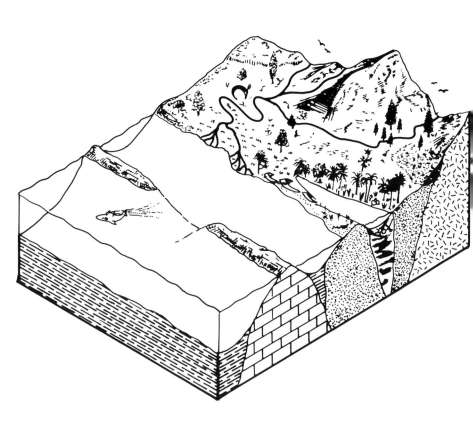

Sedimentary Rocks

Sedimentary rocks are derived from the weathering and erosion of pre-existent rocks whether igneous, metamorphic, or sedimentary, or from a combination of these. Such an origin suggests a broad spectrum of composition for the sedimentary rock class.

Rocks formed by sedimentary processes develop at or near the surface of the earth. Two major sub-classes are recognized: (1) clastic-type sedimentary rocks and (2) chemical precipitates. A less abundant, though economically important organic plant accumulation such as peat, lignite and coal also occurs, as well as fossil animal deposits such as those made up of skeletons of invertebrates.

The clastic sedimentary rocks are derived from the erosion or wearing away of pre-existent rocks. The small and large derived particles are transported by water, wind or ice to the site of deposition. The type of transporting agent to a large extent determines the texture and structure of the rocks formed. Ice is a very poor sorter of rock materials. As it melts and retreats it leaves behind aggregates of fine and coarse particle sizes all heterogeneously arranged. If this sedimentary material is buried, consolidated, and indurated, the rock product formed is called tillite.

Wind is at the other extreme and is a very good sorting agent. Only the finer particles are transported and deposited by this medium and even these are well-sorted by the wind currents. Fine-grained sands, silts and even finer clay particles are part of the sediment load which is deposited. The burial, compaction and induration of these fine-grained sediments produces sandstones, siltstones, and shales. The massive cliffs of Zion National Park, Utah are a striking example of wind deposited sandstones with their conspicuous internal cross-layered structures.

Stream waters, one of the main erosional and transporting media, mechanically erode particles of rock, carry them in suspension and as bed load, and redeposit them in the same or a different environment. As an example, it has been estimated that 500,000,000 tons of dissolved and clastic material are carried into the Gulf of Mexico by the mighty Mississippi River each year.

A stream is potentially able to carry in suspension and traction an amount of sediment equivalent to the energy available to stream flow. The greater the velocity of the stream, the larger the amount of sediment transported and also the larger the particle size carried and later deposited. Sediments are classified according to particle size into gravels and breccias of greater than 2 mm in diameter, sand 2 mm down to 1/16 mm, silts 1/16 - 1/256 mm, and as clay for those which are less than 1/256 mm in diameter. These particle sizes are the major bases for the classification of clastic sedimentary rocks. The particle size also suggests an energy environment for the sedimentary deposits.

Lake waters are also very important agents in the formation of sedimentary rocks. Within lake basins large quantities of fine-grained sediments can be deposited through streams carrying the sediments into the lakes and these waters dispersing the particles across the lake floors. Similarly in the oceans, sediments from the land through stream transport and also from wave erosion and current transport result in extensive clastic deposits.

Large quantities of the chemical weathering products of rocks are transported in solution in water. These are not mechanical or visible particles, but dissolved material in the water and invisible to the eye. They are part of the surface waters of the earth. Material in solution in stream waters is estimated at about 30% of the stream load. If this applies to the Mississippi River, about 150,000,000 tons per year of dissolved matter is carried to the Gulf of Mexico each year. The total probably exceeds this amount. To gain a better grasp of the quantity of this dissolved material in the rivers, it would be the equivalent of 1,500,000 freight cars carrying 100 tons each year or about 4,000 per day, or 166 per hour passing a set point near the mouth of the river and chemically adding this large quantity to the waters of the Gulf. It is obvious that the amount of materials in solution transported by the rivers of the world is prodigious.

In both lacustrine and marine environments chemical precipitates can form. Witness the extensive marine limestone of the mountains and plains area and the great saline deposits of New Mexico, Kansas, Michigan, New York and elsewhere–the result of both extensive chemical and evaporative processes in ancient seaways.

Organic and chemical sedimentary rocks do not lend themselves to the ready classification of the clastic sub-class. In many classifications the type of origin is often used. For example, the minerals calcite and dolomite are chemical precipitates of calcium and magnesium carbonates. These may result from direct chemical precipitation from water or be precipitated by organic processes in a marine environment.

Chert which is a locally abundant siliceous rock, may result from the accumulation of great quantities of siliceous shells or be a chemical precipitate from water. In some cases, large calcareous skeletons, such as shells, accumulate to form coquina, a type of sedimentary rock.

A third, and locally economically important sub-class of chemical precipitates are the evaporites or salt deposits. These are the residues from the evaporation of sea or lake waters. They consist mainly of sodium chloride, or salt, gypsum and anhydrite. Much rarer are the trona, sodium bicarbonate deposits, and the borates which include borax, ulexite and other borax-bearing bodies.

Peat and lignite are formed from the compaction and chemical changes on organic accumulations of plant materials. These are plant fragments which have been uniquely preserved from erosion and decay in special sedimentary environments such as swamps, some lagoons, and even some unusual stream areas.

The accumulation of sediments, chemical precipitates, organic deposits and evaporites by deposition is but a first stage of many which result in the final product we call a sedimentary rock. Burial, compaction, dehydration, induration, diagenetic changes and other processes are involved before a sediment can become a rock we can heft and classify.

Table 18 is a classification of sedimentary rock types. It is based on texture, cement, and composition.

TABLE 18. CLASSIFICATION OF COMMON SEDIMENTARY ROCKS

TEXTURE (Size & Structure of grains & particles composing the rock)	CEMENT (Composition of material holding grains together)	CLASTIC ROCKS — Quartz &/or Chert	CLASTIC ROCKS — Quartz, Chert, Mica, Rock Fragments	CLASTIC ROCKS — Quartz & Feldspars Some Clay	CHEMICAL ROCKS — Silica gel	CHEMICAL ROCKS — Calcite or Dolomite	CHEMICAL ROCKS — Gypsum	CHEMICAL ROCKS — Halite	ORGANIC ROCKS — Organic Materials	ORGANIC ROCKS — Carbon
CRYSTALLINE	N/A					Crystalline Limestone/Dolomite	Selenite	Rock salt/Halite		
COARSE, Grain size >2 mm	Silica	Quartz/Chert Conglomerate	Breccia/Fan Conglomerate	Breccia/Fan Conglomerate						
COARSE, Grain size >2 mm	Calcite & Other Carbonates	Quartz/Chert Conglomerate	Breccia/Fan Conglomerate	Breccia/Fan Conglomerate						
COARSE, Grain size >2 mm	None	Gravel	Tillite/Gravel	Tillite/Gravel						
MEDIUM, Grain or Crystal 2 mm–1/16 mm	Silica	Clean Sandstone	Argenite/Graywacke	Arkose Sandstone		Siliceous Limestone/Dol.	Selenite/Gypsum	Rock salt/Halite		
MEDIUM, Grain or Crystal 2 mm–1/16 mm	Calcite & Other Carbonates	Calc. or Dolo. Sandstone	Argenite/Graywacke	Cal. or Dolo. Arkose Sandstone		Dolomite/Limestone				
MEDIUM, Grain or Crystal 2 mm–1/16 mm	None	Clean Sandstone/Sand	Argenite/Graywacke/Sand	Arkose Sandstone/Sand		Dolomite/Limestone	Selenite/Gypsum	Rock salt/Halite		
FINE & VERY FINE, Grain or Crystal <1/16 mm	Silica	Siltstone/Claystone	Siltstone/Claystone		Opal/Chert	Porcellanite/Silic. Limestone	Anhydrite			
FINE & VERY FINE, Grain or Crystal <1/16 mm	Calcite & Other Carbonates	Calcareous or Dolomitic/Siltstone/Claystone/Marly Siltstone			Cal. or Dol. Chert	Micrite/Limestone/Dol.				
FINE & VERY FINE, Grain or Crystal <1/16 mm	None	Siltstone/Claystone/Mud			Opal/Chert	Micrite/Fine-grain Limestone	Anhydrite	Rock salt/Halite		Cannel coal
Open Fibrous Fossil Remains	N/A					Fossiliferous Limestone; COQUINA			PEAT; COQINA (with carbonate cement)	
Dense	N/A					(see fine & very fine grain or crystal size)	Anhydrite	Rock salt/Halite		COAL: Lignite (compact) Anthracite (very dense)

Clastic Sedimentary Rocks

Pebbly Sandstone: The pinkish gray sandstone consist of fine-grained quartz grains enclosing clasts of small quartz pebbles and chert. Bedding is faintly visible and results from grain size differences in the rock. The environment of original deposition must have been one of moderate energy for the streams carrying the material to the site of deposition.

Gritty Sandstone: Medium to coarse-grained quartz and chert grains, with fairly uniform sorting, and of subrounded to subangular shape make up this gritty, pinkish gray sandstone. The rock has matrix around many of the grains and approaches the graywacke rock type classification.

Fine-grained Sandstone: Composed essentially of fine-grained quartz grains, uniformly oriented, and with minor iron oxide along small bedding planes, this clean sandstone, or arenite, is a not uncommon mature sandstone. Bedding or layering of the specimen parallels the page.

Graywacke: Composed of fine to medium grains this dirty appearing, medium-to-dark gray graywacke consists of chert, rock fragments, minor quartz and a clayey matrix of finer-grained materials. Unlike true arenite, or clean sandstone, clay-like particles fill the pore space of the rock.

PEBBLY SANDSTONE

GRITTY SANDSTONE

FINE-GRAINED SANDSTONE

GRAYWACKE

5 cm

Clastic Sedimentary Rocks

Laminated Siltstone: Composed of grains with a gritty feel , finely-bedded, and light gray in color, the rock is representative of a siltstone. The fine scale bedding permits the use of "laminated" to be added to the rock name. Mineral composition is not visible by the unaided eye, but quartz and clay are probable components.

Laminated Shale: Extremely fine-grained mineral particles make up this laminated shale. The bedding is made visible by the presence of hydrocarbons (kerogen) along the bedding planes and in the mineral matrix. The shale is appropriately called a laminated oil shale because of its high kerogen content. Much research has been done to effect an economic recovery of oil from these rocks, of which the higher content shales carry as much as 25-100 gallons per ton.

Shale: Pinkish gray with local limonitic overtones, the shale shows good bedding parallel to the page and has well-developed parting on the bedding planes. The mineral composition is not recognizable in the hand specimen with the grain size less than 1/256 mm diameter. The average shale has about 30 percent quartz, about the same in clay minerals, and the other third or so in other minerals including carbonates, feldspar, iron oxides and organic matter.

Siltstone (Ripple Marked): This fine-grained clastic rock has particle sizes less than 1/16 mm and down to the clay size of 1/256 mm diameter. Siltstone, shale, and claystone are the most abundant of the sedimentary rocks. This rock is brownish red and part of the red-bed series of the Colorado Plateau. Mineral composition cannot be identified in the hand specimen, but the grittiness suggests quartz intermixed with

LAMINATED SILTSTONE

LAMINATED (Oil) SHALE

SHALE

5 cm

RIPPLE MARKED SILTSTONE

Fossiliferous Chemical Precipitates

Fossiliferous Limestones: The upper two rocks of the page are both chemical precipitates of calcium carbonate making up the rock limestone. The abundance of fossil material in each requires the use of fossiliferous limestone as proper names. Color is another modifier, and even the fossil genera could be added for more specific description. As an example, the upper sample would be called a "grayish blue, crinoidal, fine-grained limestone." Some might classify it as an *encrinite* because of the abundance of the fossil crinoids enclosed in the rock.

Diatomite: White, relatively light in weight and with the appearance of chalk, this earth material has a harsh feel. Under the microscope very small siliceous skeletons of unicellular plants called diatoms are visible. These have multitudes of intricate shapes and forms. Though made of such very fine-grained skeletons, diatomaceous beds may be measured in hundreds of feet in thickness. The purer diatomaceous earth is used as a polishing ingredient and as a filter for fluids.

Coquina: This cream-colored clastic limestone consists almost entirely of fossils, mainly shells, both whole and fragmented. It is partly cemented by calcite. Transportation distance of the shells must have been minimal as they are mainly whole in appearance and are well preserved. Those rocks made up of fine-sized shell debris are named *microcoquina*.

FOSSILIFEROUS LIMESTONE

(CRINOIDAL)

(BRACHIOPODS)

DIATOMITE

5 cm

COQUINA

Chemical Precipitates

Oolitic and Pisolitic Limestones: The terms oolites and pisolites refer to small, somewhat spherical particles mainly found in limestones and dolomites. Oolites are smaller, with diameter about 0.5-1.0 mm and larger. Pisolites are larger than 2.0 mm in diameter. The shape of individual oolites ranges from spherical to flattened ovoids. Internally, under the microscope, the structure is either radial or concentric and often both. The center, or core, may be of different composition than the oolites. Calcareous oolites are far more abundant than others of silica, hematite and dolomite. A suggested origin is an inorganic precipitation in turbulent waters, with the grains rolling with the current and gradually accumulating more and more layers of the precipitate.

Four rock samples made up of oolites and pisolites are shown.

Gray white oolite of calcite composition.

Medium gray medium grain pisolitic limestone.

Dark gray oolitic limestone.

Oolitic hematite. This rock type has been a major source of iron ore from ancient beds in the southern Appalachian Mountains of the United States.

OOLITIC LIMESTONE

PISOLITIC LIMESTONE

OOLITIC LIMESTONE

OOLITIC HEMATITE

5 cm

Chemical Precipitates

Lithographic Limestone: This compact, very fine and even-grained, tannish cream rock is called a lithographic limestone. The name comes from the use of such fine-grained stone in older times in making lithographic plates for printing. Easily etched by acid, and with uniform grain size, excellent printing results could be achieved. Note the conchoidal fracture, a sign of the uniformity and fine-grained character of the rock. It is obviously reactive to hydrochloric acid.

Chert: Brownish gray, slightly banded, and of cryptocrystalline silica, and very hard, the rock is classified as a chert. Fracture is conchoidal to splintery. Theories proposed for the origin of cherts include: 1) chemical sediment precipitate, 2) volcanic activity and silica release to ocean waters and later deposition, 3) normal sediment from activity of silica-secreting organisms and precipitation as a sediment.

Crystalline Limestone: Medium to coarse-grained, tannish cream color, and somewhat uniform character of texture are features of this crystalline limestone. Normally, a limestone is fine-grained but during diagenesis, or those chemical-physical changes which occur in the time interval between sediment and true rock formation, changes occur from original compaction. Crystallization of colloids and textural changes, including replacement, occur which bring about grain size changes in some limestones. This increase in grain size may, in some cases, result in a marble-like rock such as shown.

Phosphatic Limestone (Phosphorite): This dark gray, well-bedded, fine-grained, phosphate-bearing limestone is a product of the precipitation of apatite from upwelling cold waters when these encounter the warmer shelf waters where the solubility is 30 times less. The apatite precipitates in this environment and forms bedded phosphorites of generally considerable extent. Some mineral associates are chert, black shale, limestone and sandstone.

LITHOGRAPHIC LIMESTONE

CHERT

PHOSPHATIC LIMESTONE

CRYSTALLINE LIMESTONE

5 cm

Chemical Precipitates

Dolomite (Dolostone): Finely crystalline, of dull luster, non-reactive to acid except when warm, and gray color, this rock looks like a limestone but is a dolomite. Many such rocks have been mapped as limestones because a few rudimentary tests were neglected and the general appearance accepted as limestone. The term dolostone is applied to these rocks containing more than 50% carbonates with a high content of dolomite. They are rocks resulting from the dolomitization of limestone. The process may occur during or soon after sedimentation or at a later time when solid rock is present. It may also result from some types of hydrothermal (hot water) alteration near plutonic igneous rocks. The dolostone resembles limestone in general, but in detail distinct differences exist between them.

Algal Ball Limestone: With ball-like inclusions to almost an inch in diameter, some spherical, others ovoid, and in a matrix of fine-grained calcite, this algal-ball limestone is a distinctive brown rock. The individual "balls" show concentric internal structures. The algae apparently had a central core object made up of shell or a rock fragment about which the algal filaments grew. Calcium carbonate deposited around the fragment and the filaments and the ball-like structure formed. In a massive state, this unusual limestone is a good source rock for facing stones of buildings.

Calcareous Tufa: Of limited areal extent, calcareous tufa, through deposition from springs and surface waters, is of sporadic accumulation in the surface environment. Porous in structure, fine-grained in texture, white to dull gray to brown in color, the deposits may occur as terraces around spring orifices as in Yellowstone Park, Beowave, Nevada and elsewhere. Characteristics of the calcareous tufa are exemplified by the sample shown. The rock is vigorously reactive to hydrochloric acid.

Siliceous Sinter: As for calcareous tufa, so for siliceous sinter in its distribution pattern. It, too, is of sporadic occurrences and is almost always associated with silica-bearing spring waters where it forms irregular masses and terraces. The structure of the rock is one of irregular porosity, its grain size very fine-grained and its color white to dull brownish, all are shown in the photographic illustration.

DOLOMITE

ALGAL BALL LIMESTONE

SILICEOUS SINTER

CALCAREOUS TUFA

5 cm

Evaporites

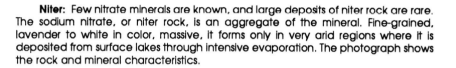

Niter: Few nitrate minerals are known, and large deposits of niter rock are rare. The sodium nitrate, or niter rock, is an aggregate of the mineral. Fine-grained, lavender to white in color, massive, it forms only in very arid regions where it is deposited from surface lakes through intensive evaporation. The photograph shows the rock and mineral characteristics.

Anhydrite: As a rock type anhydrite is generally of limited areal extent. It is a common associate with salt and gypsum. The mineral forms the mono-mineralic rock. It resembles gypsum ($CaSO_4$ $2H_2O$) but is harder and heavier and lacks water in its composition ($CaSO_4$). The mineral is a precipitate from water through the evaporative process.

Salt (Halite): This monomineralic rock, here shown, has all the properties of its mineral, halite, including taste, feel, cleavage and hardness. It forms extensive layered deposits by the evaporation of sea water in enclosed basins. Unusually large, thumb-like masses, thousands of feet in height have forcefully "intruded" and are now enclosed in younger sediments in structures called salt domes in the Gulf Coast of the United States. Salt also forms the core and is the origin of large anticlinal (upfolds) in Western Colorado and Eastern Utah. These and the layered salt of Kansas, Michigan, Ohio, and New York are mined for their salt content, some containing valuable potash salts like sylvite and carnallite.

Gypsum: The mineral gypsum in aggregate forms sedimentary beds of wide areal extent. Well-layered, often thin to massive-bedded, and in thickness to more than 100 feet, but generally much less, the monomineralic rock has the properties of its mineral component. It is usually white to gray, finely crystallie, may show fine to coarse-layering, sometimes intricately folded, and is easily scratched by the finger-nail. It finds extensive use in the building trade, being a major component of plaster and sheet rock.

NITER

ANHYDRITE

GYPSUM

ROCK SALT

5 cm

Organic Deposits–Terrestrial

Peat: Chocolate colored, spongy, porous, fibrous, and shreds of organic matter such as grasses, mosses, and woody plants are the major features of peat. The material usually accumulates in lakes, swamps, lagoons, even in deltas of rivers. At many places in the world, peat is mined on a small scale from peat bogs as a source of energy for cooking and heating. It burns, when dried, with much smoke but has poor heat output.

Lignite: Known as brown coal, lignite represents a more coalified product from the burial and further maturation of peat. It retains the wood and some other organic structures from the original accumulations. Of brown to blackish color, it lacks the high porosity of peat. On burning it produces much smoke and little heat. One variety, known as jet, is well known for its polishing characteristics and has a hardness and shiny luster quite unlike the normal lignite.

Bituminous Coal: Black, with banded structure and with luster varying for individual bands from dull to glossy and texture from finely granular to homogeneous, bituminous coal occurs in layered units of moderate to broad expanses. Thickness ranges from inches to many tens of feet. The coal represents an increase in coalification from lignite toward anthracite, the hard coal of high carbon content.

Cannel Coal: Made up of spores and wood particles and with physical properties of conchoidal fracture, dull luster, non-banded, and of black color, cannel coal is the end product of transported, deposited, buried, compacted, and chemically altered organic material. It is of more limited extent horizontally and vertically than bituminous coal. It burns well.

PEAT

LIGNITE

BITUMINOUS COAL

CANNEL COAL

5 cm

EXTRA–TERRESTRIAL ROCKS

Tektites

Tektites are an unusual rock type, of relatively rare occurrence and of high collecting value because of shape, surface texture and rarity. O'Keefe defines them as "pieces of natural glass, of size ranging from specks barely visible to the naked eye to blocks the size of a football." Generally, these rock masses and fragments are black, but may be green or even yellow. Depending on the iron content, the specific gravity ranges from 2.2 to 2.8. The latter is of bottle green color. As a glass its Mohs Scale of hardness is 6-7. Tektites have a very low water content, as well as low abundance of volatile substances. Chemically they approach the composition of intermediate igneous rocks, but have higher silica content. A sandstone enriched in mafic minerals would more closely approach the chemical content of these unusual rocks. Compositions in weight percent oxides show the following ranges: SiO_2, 48-85%; Al_2O_3, 8-18%; FeO, 1.4-11%, MgO, 0.4-28%; CaO, 0.3-10%; Na_2O, 0.3-3.9%; K_2O 1.3-5.8%, TiO_2, -0.3-1.1%.

Internally the tektites are homogeneous glasses, though some varieties consist of small glass particles welded together. Small bubble cavities and glassy inclusions are most common. Microtektites consist of small glass spheres generally less than 1mm in diameter. Rarely some tektites contain detrital fragments. Others are reported to contain the very high pressure-temperature quartz type known as coesite, but this is disputed. Of special interest is the presence of reported nickel-iron spherules in some of the glasses. These resemble meteoritic spherules. Surface sculpture shows conchoidal fracture, melt and flow structures, bald and smooth areas, and often deep grooving. Shapes are splash-form, spheroids, elongate, rod-like, dumb-bell, tear drop, disks, button-like, and irregular.

As unique as their appearance, texture, and composition is their widespread, yet well-defined, regional localities on the earth. These regions are known as strewn fields. They include those of Australasia, the Moldavite strewn field in Czechoslovakia, the North American Field from the Eocene beds of Texas and eastward into Georgia, an Ivory Coast African strewn field, and three minor fields: Libyan desert glass, the Aouellone crater glass, and Amerikanites. Researches suggest that each of these tektite regions corresponds to a single event in their origin and deposition. This is not to say that other areas cannot yield new finds. The Martha's Vineyard area of the eastern U.S. is an example of an unrelated find outside a recognized strewn field. Others may be present.

O'Keefe, after a review of more than 900 papers written on these unique objects, concluded that tektites as found are not a local product. They lie upon a wide variety of sediments or igneous rocks, with no genetic connections between them. The surface features of the tektites indicate they probably arrived at their sites of deposition as a result of "ballistic flight through space". Source areas may have been earthly or cosmic and the distance and speed must have been sufficiently great to have permitted the ablation-like marks, the melt-flow characteristics, and partly the shape of the individual tektites. Individual strewn fields also show specific ages based on radiometric dating of the included tektites. The Moldavites of southeastern Europe are about 15 my (million years) age, an age likewise equivalent to the glasses of the well-known large explosive crater in the Ries Basin of Germany. Is their a connection? The North American tektites are about 35 my old, the Australasian (Tasmania, Australia, Indonesia, Southeast Asia, and the Philippines) are 0.75 my old, the Ivory Coast about 0.9 my.

While the genesis of these most interesting objects is still in dispute, they do appear to have entered the earth's atmosphere at velocities approximating 11 km/second. Some came as smooth bodies (Australites) and others as rough bodies

TEKTITE VARIETIES

ELLIPSOIDAL

DISC

SPHERICAL

DUMBELLS

TEARDROP

5 cm

CYLINDRICAL

MUONG NONG

Meteorites

While meteorites at the earth's surface comprise an extremely small percent of the upper crust, evidence from other solid planetary bodies indicate that large numbers of meteorites have impacted them in the far distant past. Indeed, the stages of planetary body development may involve an early accretionary stage character-ized by buildup of its mass through uncounted millions of impacts by meteorites from very large masses to those of very small mass.

Much of the space near planetary bodies of our inner four planets has been swept clean of meteorites. Occasionally, we observe a fall and rarely recover meteoritic fragments from it. We do know that finer meteoritic particles are contin-uously falling on the earth. Estimates of quantity range to 1,000,000 tons or more of meteoritic material added to the earth's surface per year.

The extraterrestrial rocks which have been recovered have been intensively studied. The included minerals of these are identified. Mason (1962) reports at least 42 minerals known to be present in the various types of meteorites. Of these 7, or about 16%, are known to occur in terrestrial or earth rocks. While not in quantity, even copper, gold, and diamonds have been discovered in the meteoritic specimens studied.

Using mineral composition, texture and color criteria, a generally accepted classification was published in 1904. Some problems with the use of color and small veinlets present, plus imprecise determination of minerals led to revision and sim-plification of this system. Prior's work of 1920 became much more widely accepted. He recognized four major groups of meteorites based on contained minerals and internal structure: chondrites (5), achondrites (7), stony-iron (4), and iron (3). The number following the group name refers to the number of classes within the group.

The chondrites are most abundant and comprise about 85% of all types thus far collected. The chondrite structure is of particular interest. These are spheroidal mineral bodies about 1mm in diameter and usually made up of olivine or orthopyroxene. They are present in each of the five classes. The mineral olivine or orthopyroxene and some iron nickel are also present in each of the classes.

Achondrites lack the chondrite structure and are generally more coarsely crystalline. These type meteorites more closely approach the composition of earth-bound rocks. Nickel-iron is essentially absent in the group. The achondrites are relatively rare as finds possibly because they so closely resemble earth rocks.

Stony iron make up about 4 percent of the finds. These are characterized by silicate minerals in a matrix of nickel iron and vice-versa. Nickel content approximates 7-15% depending on the class within the group.

The irons make up the second most abundant group, possibly because of their ready recognition and sometimes very large size. In witnessed falls they make up about 7% of the stony types. Their characteristic feature is their high iron content. The average for all iron meteorites is about 91%, with about 9 percent nickel. Generally there is a very minor cobalt content of less than 1%.

Overall distinguishing features of the meteorites are the universal presence of nickel in the iron content. Its absence excludes the object as of meteoritic origin. A quick test is to dissolve a small amount of the iron recovered from the suspect sample in a small amount of nitric acid. Add some ammonium hydroxide to make the solution slightly basic. Filter and add an alcohol solution of dimethylglyoxine to the filtrate. A scarlet precipitate confirms nickel. Surface features such as scallops or meringue-like features on the specimen and an unusual dark color are also distinguishing features. If the surface of the specimen is polished, a characteristic lamellar intergrowth is almost always present in the iron mineral. The stony types also contain small particles of nickel-iron. This is easily recoverable from crushed fragments by a strong hand magnet. The recovered material can be tested for nickel as noted above. Mason (1962) suggests the advice of an expert on those specimens lacking the distinctive feature described.

IRON-NICKEL
METEORITE
(Toluca, Mex.)

IRON-NICKEL
METEORITE

IRON-NICKEL
METEORITE
(Glorietta, NM)

METALLIC OCTAHEDRITE
(Canyon Diablo, AZ)

IRON-NICKEL METEORITE
(Sacramento Mtns. NM)

5 cm

METEORITE IMPACT BRECCIA
(Wanapati Crater,
Ontario, Can.)

APPENDIX 1. DETERMINATION OF MINERALS BY STAINING TECHNIQUES

Mineral Staining Tests

Mineral staining tests are helpful in identifying and differentiating between minerals of similar appearance and/or other physical properties. The tests are fairly easy to administer if the outline in the following tables is carefully followed. Only those tests which use chemicals that are more easily obtained by the novice scientist were selected. The test outline is especially prepared for the non-chemist, avoiding specialized preparation techniques and solution concentration descriptions. However, since carelessly handled chemicals of any kind can be extremely dangerous, the following "DO's" and "DON'Ts" of laboratory safety procedures must be carefully observed.

NEVER

NEVER	pour water into acids, always pour acid into water.
NEVER	smell any chemicals directly from the container.
NEVER	mix chemicals just for the fun of it; follow instructions and know what you are doing.
NEVER	touch or taste any chemicals.
NEVER	heat anything in a closed container.
NEVER	heat any chemicals just for fun; follow instructions and know what you are doing.
NEVER	EVER put chemicals in containers used for human consumption. Always use appropriate containers and lable unmistakingly.

The authors are not responsible for any injury, damage, and/or death resulting from the use or misuse of chemicals and/or procedures described or mentioned in this book.

ALWAYS

ALWAYS	read instructions and know what you are doing BEFORE you start an experiment.
ALWAYS	wear adequate eye protection when performing experiments.
ALWAYS	wash your hands after handling chemicals. Immediately rinse with water any chemicals, particularly acids, off your skin and clothes.
ALWAYS	point away from you and others a test tube or bottle opening while heating it.
ALWAYS	keep your work area well ventilated; especially when you are boiling acids and bases.
ALWAYS	keep chemical containers closed when they are not in use.
ALWAYS	keep any kind of chemicals out of the reach of children.

The following example using the test for anhydrite and gypsum will show how the tables are used for the experiments. Follow the sequence from left to right. Chemicals needed must be measured first. They are not mixed as long as they are separated by a horizontal line. DO NOT MIX CHEMICALS UNLESS THE HORIZONTAL LINE SEPARATING THEM IS DISCONTINUOUS. In this example, 10.4 grams of Barium chloride and 100 milliliters of distilled water are mixed first. A little later, because the line continues a little further to the right, 0.3 grams of potassium permanganate is added to the mixture. after this is completed, the sample, indicated by the rock symbol, is immersed and treated according to instructions. 5 grams of Oxalic acid and 100 milliliters of distilled water are also mixed. However, they are always separated from the other chemicals because the horizontal line above the Oxalic acid continues through the whole MIXING AND TREATMENT sequence. A vertical dashed line indicates that the sample needs to be removed from the solution and thoroughly washed with distilled water. The arrow indicates that the sample is moved into the next procedure and/or solution; here, into the oxalic acid bath. After the sequence is completed, the sample should have the color or coating indicated in the column "COLOR CHANGE OR STAINING."

MINERAL	CHEMICALS NEEDED		MIXING & TREATMENT	COLORANCE OR STAINING
Anhydrite and Gypsum	Barium chloride	10.4 g		pink to red
	distilled water	100 ml		
	Potassium permanganate	0.3 g	Boil 2 minutes	
	Oxalic acid	5 g		
	distilled water	100 ml	Heat 2 minutes	

Sequence of procedure

Benzidine solution

conc. Hydrochloric acid	18 ml	1 ml
distilled water	100 ml	
distilled water		100 ml
Benzidine		2 g

The same sequence of procedures is true for preparing solutions used in the identification of certain carbonate minerals. As illustrated by the example of preparing a benzidine solution, hydrochloric acid and distilled water are mixed first. 1 milliliter of this solution is then mixed with 100 milliliters of distilled water. At last, 100 milliliters of that solution is finally mixed with 2 grams of benzidine.

MINERAL STAINING TEST

MINERAL	CHEMICALS NEEDED	MIXING & TREATMENT	COLORANCE OR STAINING
Calaverite	conc. Nitric acid	50 ml — Boil 1 minute	dull black
Pyrite	distilled water	50 ml	tarnish removed
Pyrrholite	conc. Hydrochloric acid	15 ml — Boil 3 minutes	tarnish removed
Chalcopyrite	Tin	4-5 grains	tarnish removed
Gold			no effect
Brucite Hydro–magnesite	Silver nitrate; distilled water	10 g; 100 ml — Heat sample first for 5 minutes, Immerse for a few minutes	brown to black coat
Anhydrite and Gypsum	Barium chloride; distilled water; Potassium permanganate; Oxalic acid; distilled water	10.4 g; 100 ml; 0.3 g; 5 g; 100 ml — Boil 2 minutes, Heat 2 minutes	pink to red
Barite	Sodium carbonate; distilled water; Potassium chromate; distilled water	Boil 3 minutes, Heat for a few minutes	yellow

Warm gently for 3 minutes:
- pinchbeck – brown
- little or no tarnish
- iridescent
- iridescent
- unattered

MINERAL	CHEMICALS NEEDED		MIXING & TREATMENT	COLORANCE OR STAINING
Ilmenite	conc. Hydrochloric acid		Boil 20 minutes	dull gray to brown coating
Scheelite & Wolframite	conc. Nitric acid	30 ml	Boil 30 minutes	canary – yellow
	conc. Hydrochloric acid	30 ml		
	distilled water	10 ml		
Bauxite	conc. Hydrochloric acid		Boil 5 min.	purplish – pink
	Sodium hydroxide	4.0 g	2 minutes	
	distilled water	100 ml		
	Alizarin red S	0.1 g	Add & tumble 1 minute – wait 4 minutes	
	distilled water	100 ml		
	5 N Acetic acid	30 ml	Add	
Cassiterite	conc. Hydrochloric acid	9 ml		dull gray coat at elemental tin
	distilled water	100 ml		
	Zinc powder	15 g	5 minutes	
Chalcocite	Ferric chloride	20 g	2 minutes	blue
	distilled water	100 ml		
Stibnite Kermesite	Potassium hydroxide	40 g	5 minutes	yellow
	distilled water	100 ml		
Sulfide Minerals	Silver nitrate	1.7 g	Solution is light sensitive Boil for 1 minute	Dull grains – Galena Grains cohere – Arsenopyrite Grey coating of Silver – Bornite, Cuprite, Chalcocite, or Niccolite No apparent change – Bornite, Pyrite, Sphalerite, Stannite or Tetrahedrite
	distilled water	100 ml	Blue or purple film – Chalcopyrite Pale orange-yellow film – Pentlandite Purplish brown film – Pyrrhotite	

MINERAL	CHEMICALS NEEDED		MIXING & TREATMENT	COLORANCE OR STAINING
Feldspathoids (fresh samples only)	conc. Phosphoric acid	85 ml	3 min. → 1 min.	deep blue – Nepheline, Sodalite or Analcime
	distilled water	15 ml		pale blue – Melilite
	Methylene blue	0.25 g		no color – Leucite
	distilled water	100 ml		
Potash Feldspar	conc. Hydrochloric acid	50 ml	Heat for 5 min.	yellow or pink
	distilled water	50 ml		
	Ethyl alcohol	30 ml	Warm gently for 15 min. / Tumble while washing	
	Safranine O	2.0 g		
	distilled water	70 ml		
Columbite –Tantalite Wolframite Scheelite Cassiterite	conc. Hydrochloric acid	25 ml	Boil 20 min. → No stain / yellow grains → Do Cassiterite staining test / Remove yellow stain	dull gray coat – Cassiterite
	conc. Nitric acid	15 ml		no coat – Columbite – Tantalite
	distilled water	60 ml		white grains – Scheelite
	Ammonia			dark grains – Wolframite

PREPARATION OF SOLUTIONS FOR CARBONATE MINERAL DETERMINATIONS

Feigl's solution

Manganese (II) sulfate, heptahydrate	11.8 g
distilled water	100 ml
silver nitrate	1 g
sodium hydroxide	0.1 g
distilled water	10 ml

Boil

Add 1-2 drops after cooling →

After 2 hours filter off precipitate and store solution in dark bottle.

Benzidine solution

conc. Hydrochloric acid	18 ml	1 ml
distilled water	100 ml	
distilled water		100 ml
Benzidine		2 g

Rhodizonic acid solution

Disodium rhodizonate	2g
distilled water	100 ml

Alizarin red S solution

conc. Hydrochloric acid	0.2 ml	100 ml
distilled water	100 ml	
Alizarin red S		0.1 g

Magneson solution

Sodium hydroxide	1 g	100 ml
distilled water	100 ml	
Magneson (P-nitrobenzene –azo – resorcinol		0.5 g

SCHEMATIC OF MINERAL STAINING TESTS FOR CARBONATE MINERALS

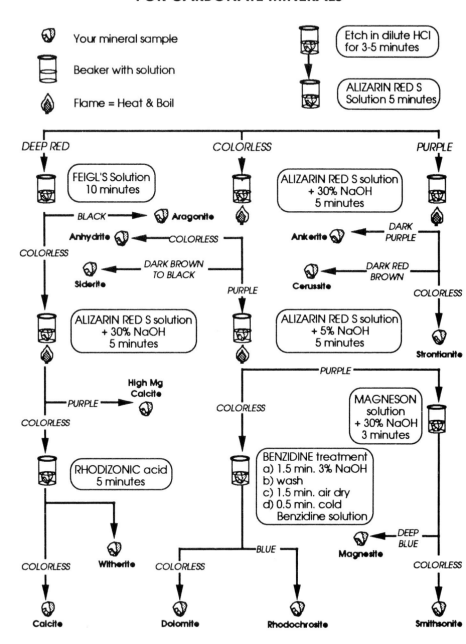

(Mod. after Warne, 1962)

APPENDIX 2. THE MINERAL COLLECTION

There is one question left. What to do with all those rocks that have accumulated in cardboard boxes over the years? You have probably read several suggestions about starting, labeling and sorting mineral collections in huge drawer cabinets out of sight. Frankly, most of us have little room for monstrous cabinets lining the wall of some hallway with a "perfect" but yet invisible rock collection. In the following pages are a few tips for a more aesthetic use of rock collections.

Sorting

The first step in any mineral or rock collection is the sorting of the specimens, separating the best pieces from the worst. Those displaying exquisite crystal structure or an unusual make-up should be kept. Small, but perfect crystals may be kept for unique gift ideas as explained below. Large amounts of pretty or unusual specimens may be used as permanent decorations in gardens or fireplaces (if you are planning on building or remodeling in the near future.) Small pieces of certain rocks and minerals may be kept to test certain cleaning and preserving techniques before using them on your "irreplaceable" specimens. But, what to do with the left-overs? Throw them away? This would be a waste of material. How about keeping them for trading purposes. Rock-hounds are always willing to trade. You may not receive 100% perfect specimens that way, but long sought samples may be obtained. Another way of giving those unwanted specimens a special purpose is to donate them to your local public schools. Teachers in those institutions will be more than happy to accommodate those specimens and donations of this kind can be tax deductible. Check your local and federal tax codes for more detail on this subject.

Display

Now that you have decided which specimens to keep, how can they be displayed? Here are some suggestions for a permanent location for those specimens. Larger amounts of brick sized samples may be incorporated into a fireplace. If you design or plan to build a new home, or try to remodel an existing home, you may look into this option. Most samples of igneous, metamorphic and sedimentary rocks are suitable for this purpose. Carbonate rocks and minerals should not be used because they may decompose with heat. Sandstone may be used, but its cementing agent (the one that holds the sand grains together) may deteriorate if heated. For safety purposes, no minerals should be used in the firebox itself, but rather on the hearth and face of the fireplace. All silicate minerals are excellently suited for this purpose.

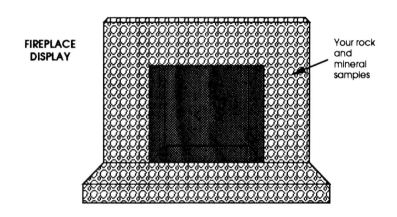

FIREPLACE DISPLAY

Your rock and mineral samples

Some minerals are not to be used for such application because they may release toxic gases. Realgar, for example, may fill your room with arsenic vapor, which obviously is a health hazard. You are safe if you avoid all ore minerals around the fireplace. Even iron pyrite may contain "harmful" substance.

Very large rock and mineral specimens can make interesting landscaping materials for your garden or porch. However, some minerals may not take too kindly to the particular climate in your area. Large gypsum crystals may do fine in a garden in St. George, Utah, but may decompose rapidly in Houma, Lousiana. Samples of iron pyrite may tarnish over a period of time in any climate. Most rock samples will do quite well, especially metamorphic and igneous ones. The samples used for landscaping purposes should be large for several reasons. A large sample is much more attractive to the eye and dishonest bypassers may resist the temptation or risk a serious hernia.

In-room displays are very attractive for mid-sized and small samples. Larger pieces may sit on the floor or on coffee tables. They may also look beautiful on bookshelves between groups of books. Bookends made out of rocks and minerals is another possibility but may require some special tools or treatment of the specimen by rock-shops. Specimens may also look very attractive if displayed on rustic looking wooden tripods. This is especially useful for samples that cannot be put on a flat surface.

Old type or letter cases from antiquated printing shops can be wall mounted and make very handsome display cases for smaller specimens. If you are talented in woodworking, you may try to tackle one of the following projects:

1) A picture or fixed window may become a very attractive display case. Install several glass shelves inside the window not wider than the sill space. On the top of the window, install a hidden fluorescent light fixture. A glass door on the inside of the window will keep dust away from your specimens. Now you can place your samples on the glass shelves. Translucent and transparent specimens are excellently suited for this purpose and may even display an array of colors through the incoming sunlight. the fluorescent light coupled to a timer or light sensor will take over at dusk. You may even enhance the appearance of your samples by using a black light tube. Fluorescent minerals will now glow in all colors.

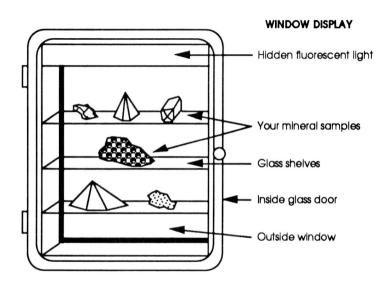

WINDOW DISPLAY

Hidden fluorescent light

Your mineral samples

Glass shelves

Inside glass door

Outside window

2) Another possibility is to build a living room coffee table with a glass top and a display case below. Fluorescent lights may be hidden around the edges. Black light fluorescent tubes may be used as a second set of lights illuminated by a special switch or mixed with regular light to enhance color. The table will not only be very attractive, but can add soft light to a room.

An alternative to this type of a table would be to make the surface of the table out of mineral specimens or have the surface contain mineral specimens. A table top can be made by manufacturing a plexiglas table-top that has mineral specimens as inclusions. This, however, takes quite a bit of expertise and may require some professional help. Relatively flat specimens may be laid like tile and a tile adhesive can be used to affix the samples to a prepared table top. The gaps can be grouted with ordinary tile grout.

Some people like the idea of displaying minerals and rocks in aquariums and terrariums. A word of caution may be given to them. Minerals may decompose in water or may give off toxins that are harmful to fish and plant-life. In general, silicate minerals are safe to use and ore minerals should be avoided. Carbonate minerals could be used, but they are likely to decompose and increase the hardness of the water dramatically. This may be fatal to certain sensitive fish species.

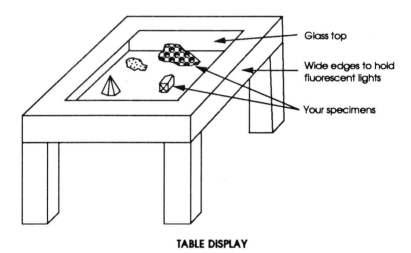

TABLE DISPLAY

3) Powerful small magnets, a good stickum substance, a painted or decorated thin steel sheet attached to the wall and minerals of appropriate size you would like to display, and voila, you have a very attractive and changeable mineral display. These make for good conversation pieces. Costs are minimal, and the effect maximal. Try it and see. A word of caution – keep the steel plate high enough on the wall to avoid the grasping fingers of children, but still in a conspicuous enough location in your home so that visitors can enjoy your prized possessions of the mineral kingdom.

We like the easy mobility of this display. Tire of one or more specimens, change them. Even a scene on the plate can be changed by repainting. Plate size is the width of the center to center line of the studs. Be sure the stickum is long lasting and does not release you specimens and let them fall to the floor. Carpet beneath the display may be an insurance.

The wall hanger display built of light wood or plastic is another method of displaying your mineral and rock specimens.

WALL HANGER DISPLAY

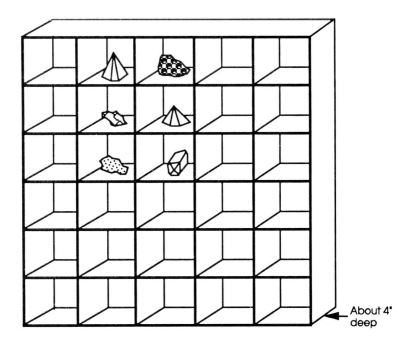

About 4"
deep

Cleaning and Preparation

A word about cleaning and preserving minerals. The best cleaning agent available is water and all specimens, unless water soluble, can be carefully cleaned in water. Harder specimens, such as quartz, can be mechanically cleaned using wire brushes, preferably the softer kind made out of bronze or copper wire. Softer minerals may be carefully cleaned with tooth brushes. Ultrasonic jewelry or pen cleaners have been proven useful for smaller hardy specimens. Minerals that cleave or fracture easily may be seriously damaged by these instruments.

Most of the iron oxide stains can be removed with hydrochloric acid. Extremely good results can be obtained by mixing oxalic acid with the hydrochloric acid. Tile cleaning or muriatic acid from hardware stores is hydrochloric acid of a useful concentration. Leave hardy specimens overnight in the acid bath. Strong acids decompose certain minerals, such as sphalerite, and may destroy a beautiful specimen. Care should be taken and cleaning techniques should be tested on some smaller samples.

Clear lacquer, brushed on or sprayed, is an excellent preservative for many minerals. Test the lacquer first on some of your less desirable samples or on the "invisible" side of your specimen to avoid unpleasant surprises. Shellac is preferable to other lacquers because it does not react with time. Certain lacquers may yellow or react with your samples over a period of several months.

Those of you who would like to know more about the cleaning and preservation of minerals are referred to Richard M. Pearl's booklet "Cleaning and Preserving Minerals" (1982), published by the Earth Science Publishing Company, P.O. Box 1815, Colorado Springs, Colorado, 80901. Many useful ideas from mineral experts all over the world are condensed here and easy to follow recipes are given.

APPENDIX 3. SOURCES FOR INFORMATION ON MINERAL–ROCK COLLECTING LOCALITIES AND MUSEUMS

As a major assistance in identifying excellent collecting localities, museums and other places of mineral and rock interest, the following agencies in the United States and Canada will be of great value. Many have brochures and maps of special interest to you. Addresses and telephone numbers are included for each agency.

STATE GEOLOGICAL SURVEYS

Geological Survey of Alabama
Box O, University Station
Tuscaloosa, Alabama 35486
(205) 349-2852

Alaska Division of Mining & Geologic Survey
794 University Ave., Basement
Fairbanks, Alaska 99709
(907) 494-7625

Arizona Geological Survey
845 North Park Ave.
Tucson, Arizona 85719
(602) 621-7906

Arkansas Geological Commission
Vardelle Parham Geology Center
3815 W. Roosevelt Road
Little Rock, Arkansas 72204
(501) 371-1488

California Division of Mines & Geology
1416 Ninth St., Room 1341
Sacramento, California 95814
(916) 445-1923

Colorado Geological Survey
1313 Sherman St., Room 715
Denver, Colorado 80203
(303) 866-2611

Connecticut Geological & Natural History Survey
State Office Building
165 Capitol Ave., Room 553
Hartford, Connecticut 06106
(203) 566-3540

Delaware Geological Survey
University of Delaware
101 Penny Hall
Newark, Delaware 19716
(302) 451-2833

Florida Bureau of Geology
903 W. Tennessee St.
Tallahassee, Florida 32304
(904) 488-4191

Georgia Geologic Survey
Room 400
19 Martin Luther King Jr. Drive SW
Atlanta, Georgia 30334
(404) 656-3214

Hawaii Division of Water & Land Development
Box 373
Honolulu, Hawaii 96809
(808) 548-7533

Idaho Geological Survey
Morrill Hall, Room 332
University of Idaho
Moscow, Idaho 83843
(208) 885-6195

Illinois Geological Survey
Natural Resources Building
615 E. Peabody Drive
Champaign, Illinois 61820
(217) 333-4747

Indiana Geological Survey
611 N. Walnut Grove
Bloomington, Indiana 47405
(812) 335-2863

Iowa Geological Survey
123 N. Capitol St.
Iowa City, Iowa 52242
(319) 335-1574

Kansas Geological Survey
1930 Constant Ave. A
West Campus, University of Kansas
Lawrence, Kansas 66046
(913) 864-3965

Kentucky Geological Survey
311 Breckinridge Hall
University of Kentucky
Lexington, Kentucky 40506-0056
(606) 257-5863

Louisiana Geological Survey
Box G, University Station
Baton Rouge, Louisiana 70893
(504) 342-6754

Maine Geological Survey
Department of Conservation
State House Station 22
Augusta, Maine 04333
(207) 289-2801

Maryland Geological Survey
2300 St. Paul St.
Baltimore, Maryland 21218
(301) 554-5503

**Massachusetts Department of Environ-
mental Quality Engineering**
Executive Office of Environmental
Affairs
100 Cambridge St.
Boston, Massachusetts 02202
(617) 727-9800

Michigan Geological Survey Division
Box 30028
Lansing, Michigan 48909
(517) 334-6923

Minnesota Geological Survey
2642 University Ave.
St. Paul, Minnesota 55114-1507
(612) 627-4780

Mississippi Bureau of Geology
Box 5348
Jackson, Mississippi 39216
(601) 354-6228

**Missouri Division of Geology & Land
Survey**
Box 250
Rolla, Missouri 65401
(314) 364-1752

Montana Bureau of Mines & Geology
Montana College of Mineral Science &
Technology
Butte, Montana 59701
(406) 496-4180

**Nebraska Conservation & Survey
Division**
113 Nebraska Hall, University of
Nebraska
Lincoln, Nebraska 68588-0517
(402) 472-3471

Nevada Bureau of Mines & Geology
University of Nevada
Reno, Nevada 89557-0088
(702) 784-6691

**New Hampshire Department of
Resources & Economic Development**
117 James Hall, University of New
Hampshire
Durham, New Hampshire 03824
(603) 862-3160

New Jersey Geological Survey
CN-029
Trenton, New Jersey 08625
(609) 292-1185

**New Mexico Bureau of Mines & Mineral
Resources**
Campus Station
Socorro, New Mexico 87801
(505) 835-5420

New York State Geological Survey
3136 Cultural Education Center
Empire State Plaza
Albany, New York 12230
(518) 474-5816

**North Carolina Department of Natural
Resources & Community Development**
Division of Land Resources, Box 27687
Raleigh, North Carolina 27611
(919) 733-3833

North Dakota Geological Survey
University Station
Grand Forks, North Dakota 58202-8156
(701)777-2231

Ohio Division of Geological Survey
Fountain Square, Building B
Columbus, Ohio 43224
(614) 265-6605

Oklahoma Geological Survey
830 Van Vleet Oval, Room 163
Norman, Oklahoma 73019
(405)325-3031

**Oregon Department of Geology &
Mineral Industries**
910 State Office Building
1400 SW 5th Ave.
Portland, Oregon 97201-5528
(503) 229-5580

Pennsylvania Bureau of Topographic & Geologic Survey
Department of Environmental Resources
Box 2357
Harrisburg, Pennsylvania 17120
(717) 787-2169

Puerto Rico Department of Natural Resources
Geological Survey Division
Box 5887
Puerta de Tierra, San Juan, P.R. 00906
(809) 724-8774

Rhode Island Statewide Planning Program
Department of Geology
University of Rhode Island
Kingston, Rhode Island 02881
(401) 792-2265

South Carolina Geological Survey
Harbison Forest Road
Columbia, South Carolina 29210
(803) 737-9440

South Dakota Geological Survey
Science Center
University of South Dakota
Vermillion, South Dakota 57069-2390
(605) 677-5227

Tennessee Division of Geology
Custom's House
701 Broadway
Nashville, Tennessee 37219-5237
(615) 742-6691

Texas Bureau of Economic Geology
University of Texas
Box X, University Station
Austin, Texas 78712-7508
(512) 471-1534

Utah Geological & Mineral Survey
606 Black Hawk Way
Salt Lake City, Utah 84108-1280
(801) 581-6831

Vermont Geological Survey
Agency of Environmental Conservation
103 South Main St.
Waterbury, Vermont 05676
(802) 244-5164

Virginia Division of Mineral Resources
Box 3667
Charlottesville, Virginia 22903
(804) 293-5121

Washington Division of Geology & Earth Resources
Department of Natural Resources
Olympia, Washington 98504
(206) 459-6372

West Virginia Geological & Economic Survey
Mont Chateau Research Center
Box 879
Morgantown, West Virginia 26507-0879
(304) 594-2331

Wisconsin Geological & Natural History Survey
3817 Mineral Point Road
Madison, Wisconsin 53705
(608) 262-1705

Wyoming Geological Survey
Box 3008, University Station
University of Wyoming
Laramie, Wyoming 82071
(307) 742-2054

U.S. GEOLOGICAL SURVEY GEOLOGICAL DIVISIONS

Headquarters, Virginia
U.S.G.S., National Center
12201 Sunrise Valley Drive
Reston, Virginia 22092
(703) 648-4000

Alaska
U.S.G.S.
4230 University Drive
Anchorage, Alaska 99508-4664
(907) 271-4398

Arizona
U.S.G.S.
2255 N. Gemini Drive
Flagstaff, Arizona 86001
(602) 527-7000

California
U.S.G.S.
345 Middlefield Road
Menlo Park, California 94025
(415) 853-8300

Denver
Denver Federal Center
Box 25046
Lakewood, Colorado 80225
(303) 236-5438

Hawaii
U.S.G.S.
Hawaiian Volcano Observatory
Hawaii National Park, Hawaii 96718
(808) 967-7328

Massachusetts
U.S.G.S.
Quissett Campus
Woods Hole, Massachusetts 02543
(617) 548-8700

Washington
U.S.G.S.
Cascades Volcano Observatory
5400 MacArthur Blvd.
Vancouver, Washington 98661
(206) 696-7693

CANADIAN PROVINCIAL SURVEYS

Alberta
Alberta Geological Survey
4445 Calgary Trail South
Edmonton, Alberta T6H 5R7
(403) 438-0555

British Columbia
Geological Survey Branch
Mineral Resources Division
Parliament Buildings
Victoria, British Columbia V8V 1X4
(604) 387-5975

Manitoba
Geological Services Branch
555-330 Graham Ave.
Winnipeg, Manitoba R3C 4E3
(204) 945-6559

New Brunswick
Department of Forests, Mines & Energy
Box 50
Bathurst, New Brunswick E2A 3Z1
(506) 546-6600

Newfoundland
Department of Mines & Energy
Box 4750
St. John's, Newfoundland A1C 5T7
(709) 576-2763

Northwest Territories
Geology Division, Northern Affairs Prog.
Box 1500
Yellowknife, N. W. Territories 04E 1H0
(403) 920-8212

Nova Scotia
Nova Scotia Department of Mines &
'Energy
Box 1087
Halifax, Nova Scotia B3J 2X1
(902) 424-4700

Ontario
Ontario Geological Survey
77 Grenville St.
Toronto, Ontario M5S 1B3
(416) 695-1283

Prince Edward Island
Department of Energy & Minerals
Box 2000
Charlottetown, Prince Edward Island
C1A 7N8
(902) 892-1094

Quebec
Ministére de l'Energie et des Res-
sources, Gouvernement du Québec
1620 Blvd. de l'Entente
Québec G1S 4N6
(418) 643- 4617

Saskatchewan
Geological Survey
Toronto Dominion Bank Building
1914 Hamilton St.
Regina, Saskatchewan S4P 4V4
(306) 787-2613

Yukon
Exploration & Geological Services
Division
200 Range Road
Whitehorse, Yukon Territory Y1A 3V1
(403) 668-5151

SUGGESTED READINGS

Bowen, W.L., 1956, *Evolution of the Igneous Rocks*, Dover Publications, New York, 110 p.

Butler, B.S., and Loughlin, G.F., 1920, The ore deposits of Utah, U.S. Geol. Surv. Prof. Paper 111, 672 p.

Chace, Fred M., 1951, Systematic Microchemical Analysis Chart, U.S. Geological Survey.

Ford, W.E., 1949, *A Textbook of Mineralogy*, John Wiley & Sons, New York, 851 p.

Gary, Margaret, McAfee, R. Jr., and Wolf, C.L. (Editors), *Glossary of Geology*, American Geological Institute, Washington D.C., 805 p. and references 52 p.

Hamilton, W.R., Woolley, A.K., and Bishop, A.C., 1974, *A Guide to Minerals, Rocks, and Fossils*, Crescent Books, New York, 320 p.

Hurlbut, Cornelius S. Jr. and Klein, Cornelius, 1977, *Manual of Mineralogy*, 19th Ed., John Wiley & Sons, New York.

Lepedes, D.N. (Editor in chief), 1978, McGraw-Hill *Encyclopedia of the Geological Sciences*, McGraw-Hill, Inc., 915 p.

Mason, B., 1962, *Meteorites*, John Wiley & Sons, New York, 274 p.

Micro-Chem Industries, 550 South Loader Ave., Pleasant Grove, UT 84602.

O'Keefe, J.A., 1976, *Tektites and Their Origin*, Elsevier Scientific Publishing Company, N.Y., 254 p.

Pearl, Richard M., 1982, *Cleaning and Preserving Minerals*, Earth Science Publishing Company, Colorado Springs, Colo.

Reid, W.P., 1969, "Mineral Staining Tests", Colorado School of Mines, Mineral Industries Bulletin 12, 20 p.

Ranson, Jay Ellis, 1962, *The Rock Hunters' Range Guide*, Harper and Brothers, N.Y., 213 p.

Short, M.N., 1940, Microscopic Determination of the Ore Minerals, 2nd Ed., U.S. Geological Survey Bulletin 914, 314 p.

Spock, L.E., 1962, *Guide to the Study of Rocks*, Harper and Brothers, N.Y., 298 p.

Tabor, D., Hardness Scale, 1954.

Tennissen, A.C., 1983, *Nature of Earth Materials*, Prentice Hall, Englewood Clifs, N.J., 415 p.

MINERAL AND ROCK INDEX

GENERAL SUBJECT INDEX

ERRATA

Page 7   Continue the last line with:
             prisms three equal-length axes intersect the prism faces as shown
             (Fig. 4).

Page 13   footnote to B-Table:
             V = valence

Page 14   footnote to C-Table:
             IR = Ionic Radius
             Insert below B on page 13

Page 26   Figure 19 spelling should read "fluorite."

Page 171   Replace SANADINE with
             SANIDINE

Page 230   Continue the last line with:
             and likely source is melted and later glassified moon material formed
             from meteorite impacts. These small masses escaped the moon,
             entered the earth's atmosphere at different times and places, and
             impacted the earth's surface.